D0622023

RECENT
Vertebrate
Carcasses
AND THEIR
Paleobiological
Implications

RECENT Vertebrate
Paleobiological

FRONTISPIECE. Carcass assemblage of horses that died of starvation, near Kras-lawka, west of Dunaburg; these horses were turned out into the fields after the Russian Revolution broke out.

Carcasses AND THEIR Implications

Johannes Weigelt

Translated by Judith Schaefer

Foreword by Anna K. Behrensmeyer
and Catherine Badgley

THE UNIVERSITY OF CHICAGO PRESS
CHICAGO AND LONDON

JOHANNES WEIGELT (1890–1948) was professor of geology and paleontology at Martin Luther University, Halle-Wittenberg. He wrote more than 150 papers and founded the Museum for Earth Science of Central Germany at Halle (now in the German Democratic Republic).

This work is a translation of *Rezente Wirbeltierleichen und ihre paläobiologische Bedeutung* by Johannes Weigelt, published in Leipzig by Verlag von Max Weg (1927).

The University of Chicago Press, Chicago 60637
The University of Chicago Press, Ltd., London
© 1989 by The University of Chicago
All rights reserved. Published 1989
Printed in the United States of America

98 97 96 95 94 93 92 91 90 89   5 4 3 2 1

Library of Congress Cataloging-in-Publication Date

Weigelt, Johannes, 1890–1948.
   [Rezente Wirbeltierleichen und ihre paläobiologische Bedeutung. English]
   Recent vertebrate carcasses and their paleobiological implications / Johannes Weigelt ; translated by Judith Schaefer ; foreword by Anna K. Behrensmeyer and Catherine Badgley.
       p.   cm.
   Translation of: Rezente Wirbeltierleichen und ihre paläobiologische Bedeutung.
   Bibliography: p.
   Includes index.
   ISBN 0-226-88166-0 (alk. paper). — ISBN 0-226-88167-9 (pbk. : alk. paper)
   1. Vertebrates, Fossil. 2. Paleoecology. 3. Vertebrates, Fossil—Texas—Smithers Lake. 4. Paleoecology—Texas—Smithers Lake.
I. Title.
QE841.W413   1989   89-4698
566—dc19   CIP

∞ The paper used in this publication meets the minimum requirements of the American National Standard for Information Sciences—Permanence of Paper for Printed Library Materials, ANSI Z39.48-1984.

Contents

Foreword

Anna K. Behrensmeyer and Catherine Badgley

Picture the tranquility of the Texas coastal plain in the mid-1920s—wide marshes, broken clouds, the occasional lowing of cattle, the animated scurrying of sandpipers, the hum of mosquitoes. Over the mudflats and beaches wanders an energetic German scientist, absorbed in recording details about the death and burial of cows, birds, alligators, and fishes. Johannes Weigelt's research activities undoubtedly raised the eyebrows of local ranchers and townspeople. We can only wonder today at the impressions he made upon the living residents of the coastal regions in his enthusiastic search for information about the dead.

Weigelt went to Texas on a mission: to gather observations about organic preservation in modern depositional environments. The larger purpose of this mission was to link processes of mortality, transport, and burial with patterns of fossil accumulation in the vertebrate record. He returned to Germany after sixteen months of intensive field work in Louisiana, Oklahoma, and Texas, and in short order produced the original version of this book—the first major work on vertebrate taphonomy, published in 1927. The volume has been an important source of information about natural processes affecting vertebrate remains for readers of German, but its impact on most members of the English-speaking paleontological community has heretofore been limited.

In this welcome translation, Weigelt leads us on a fascinating excursion through his observations on modes of death, decomposition, and burial among the vertebrates. Judith Schaefer's translation vividly transmits Weigelt's personable writing style and excitement with his subject. Weigelt wrote in the manner of a nineteenth-century naturalist, with an anecdotal format and digressions that are not characteristic of scientific writing in the late twentieth century. His style, however, effectively involves the reader in examination of the sometimes distasteful processes of death and decay. Weigelt occasionally provides glimpses of his personal reactions to these subjects, as when he makes a plea for control of oil spills, which caused the deaths of many waterfowl that he observed. His unsqueamish yet sensitive accounts convey a strong impression of the man as well as of his work. Undoubtedly, it would have been memorable to spend a day in the field with Weigelt, sharing his splendid powers of observation and his recognition of the diversity of the natural processes that transform organic remains into fossils.

Weigelt was the first naturalist to mount a full-scale research effort to document processes of vertebrate death, decay, disarticulation, transport, and burial, and to determine their relevance to fossil preservation. He referred to the study of modern phenomena as analogues for the past, or actualism, as his "ontological method." He called the application of this method to processes leading to the embedding and preservation of organic remains "Biostratinomie." His research provided early ground-work for the field of taphonomy, formally named by the Russian paleon-tologist I.A. Efremov in 1940. As originally conceived by Weigelt, "Bio-stratinomie" is close to "taphonomy" in meaning, but later usage restricted biostratinomy to transformations of organic remains from death through burial and excluded important later processes such as diagenesis. Taph-onomy is now the more inclusive term.

Weigelt possessed an unusual scientific mind that moved among widely differing fields with ease, and he used what he learned in one field to spark new insights in another. Taphonomy was only one of many research in-terests. He began his career as an economic geologist, working with the copper-bearing deposits (the Permian Kupferschiefer) in Germany. Here he became familiar with the term "Lagerstätten," which refers to concen-trated bodies of ore, coal, or evaporites. He later recast this term in a paleontological context as "Fossil Lagerstätten" to refer to large concen-trations of especially well-preserved fossils. The term and the focus pro-vided on processes responsible for creating massed organic remains have become widely used in paleontology.

Stressing the importance of a "documented, accurate understanding of specimens used for comparison" with the fossil record, Weigelt an-ticipated how much actualistic taphonomy could contribute to paleo-ecological studies. In this book he presents elaborate accounts of the fates of carcasses under a range of natural circumstances in both marine and continental environments, combining his own observations, communi-cations from his scientific colleagues, published references from a wide range of fields including forensics, and occasional anecdotes from un-specified sources. Although his presentation of circumstances of death and decay of individual cows and gars may seem overly detailed, it becomes clear that these examples represent potential analogues for specific paleontological cases. Thus, he makes readers repeatedly aware of connections between the past and the present.

In considering the causes of fossil lagerstätten, Weigelt held that much of the fossil record represents catastrophic mortality due to infrequent events of unusual violence, such as severe storms and floods. He argued that the confluence of unusual biological and geological processes makes substantial contributions to the fossil record. These processes are critical to the preservation of soft body parts, massive bonebeds, and articulated skeletons. He was impressed with the effects of panic and stampede as causes of mass deaths among herding animals and saw that the trapping effects of mud on Texan motorcars would have been equally effective for dinosaurs. In addition, he was concerned with recognizing circumstances in which slower, attritional morality and accumulation could lead to fossilization. He specified sedimentary environments, such as abandoned channels, in which vertebrate remains are likely to be concentrated through repeated accidents, such as mud-trapping.

The importance of Weigelt's studies on the Gulf Coast to his own paleontological work was apparent when he applied his observations to the "Leichenfelder" or "dying fields" of superbly preserved vertebrates in the Eocene Geiseltal lignites of East Germany. Based on his observation of superimposed mass deaths at Smither's Lake, Texas and elsewhere, he predicted that the Eocene deposits would produce more than one interval with concentrated skeletal remains. His student E. Voigt's continuing excavations proved this prediction correct. Weigelt used his records of carcass positions in different modern situations to help interpret mode of death and interment for individual carcasses in the Eocene deposits. He must have experienced considerable excitement and satisfaction in discovering analogues in the patterns of death and burial of vertebrates across forty million years.

Weigelt's observations and the issues he raises should seem very current to readers of recent publications in vertebrate taphonomy. His factual records and interpretations indicate the value of further empirical studies of recent mortality, disarticulation, hydrodynamic transport of bones, decomposition of carcasses, spatial characteristics of skeletal assemblages on land surfaces, and circumstances of burial. Readers also will find a wealth of ideas that can act as a source of taphonomic hypotheses for future research on fossil assemblages. His work highlights the role of ecological as well as sedimentary processes in fossil preservation, and the need for further research in characterizing such processes in particular depositional environments. The book also raises questions worthy of renewed investigation: How much of the fossil record is a product of unusual events, and what are the consequences of these events for paleoecological and evolutionary studies?

We can easily imagine how excited Weigelt would have been to see how the role of taphonomy is expanding along with the vast increase in information from the fossil record, and with new goals for the reconstruction of faunas and environments through time. There is no tribute he would have appreciated more than to know that what he learned is helping to catalyze new efforts in taphonomic research. This translation can now provide a much larger audience with ideas to ponder, hypotheses to test, and additional observations to make that will further enrich our understanding of taphonomic processes and their effects on the fossil record.

Anna K. Behrensmeyer
Washington, DC

Catherine Badgley
Ann Arbor, MI

5 May 1988

Translator's Note

In 1962, a memorial tribute to Johannes Weigelt was published by the Geologisches Staatsinstitut in Hamburg.* The author of this tribute, Professor Ehrhard Voigt, wrote: "Weigelt's authoritative works . . . place him in the company of those scientists who have provided the inspirational foundation for geology and paleontology." "Brilliant men cannot be measured by the same yardstick as others. . . . We must grant a certain subjectivity to a dynamic nature like Weigelt's, for without that quality, his creativity could not have attained its full potential. . . . Content was more important to him than form, and he had no use for the niceties of outward appearance. . . . The spoken word fascinated him more than the printed."

It is not surprising that a book written by such a man might need editorial attention, and this translation has provided an opportunity to make some minor changes in the original text. These changes have been made with great care, so that readers will be able to refer to this translation when using sources citing the original German edition. Therefore none of the figures, plates, notes, or references has been renumbered, although the plate numbers have been changed from roman to arabic numerals. The original book consisted of large parts, each comprising many sections. Here the parts are called chapters, a usage to which modern readers are more accustomed.

In chapter 4, section 4, I have rearranged about four pages of material to help the text flow more smoothly; no information has been omitted. Lesser changes include the occasional deletion of a redundant word or phrase and the correction of obvious typographical errors.

Two kinds of information have been added. Where necessary, scientific names have been updated, and the new terminology appears in brackets after the names Weigelt used. I thank Donald E. Russell of the Institute of Paleontology, National Museum of Natural History, Paris, for his generous and indispensable help in reviewing scientific names and providing

*Ehrhard Voigt, "Johannes Weigelt als Paläontologe," *Mitteilungen aus dem Geologischen Staatsinstitut, in Hamburg* 31 (1962): 26-50.

many current ones. In addition, current names of countries and localities that Weigelt mentions have been added, again in brackets following the place names Weigelt used.

Because the original art for the illustrations is no longer available, the illustrations in this translation are reproduced directly from the German edition; consequently, the line drawings in particular are not always as clear as we had hoped they would be.

I would like to express my deepest gratitude to Karl Hirsch, associate at the University of Colorado Museum, for proposing that I attempt this translation; for his careful, word-for-word, critical review of the manuscript; and for his unstinting support of my efforts as a translator. Further thanks go to Peter Robinson, Professor of Natural History and Curator of Geology at the University of Colorado, who not only encouraged my work but provided access to a computer. Special thanks go to Hartmut Haubold, Professor of Natural Science at Martin Luther University in Halle an der Saale, German Democratic Republic, who gave me a tour of the Geiseltal Museum (where Johannes Weigelt worked), provided a second copy of the German edition from which to reproduce the plates, helped locate modern place names, and directed me to the quotation that opens this note.

I acknowledge with thanks the support and encouragement of Anna K. Behrensmeyer, research curator in the Department of Paleobiology, National Museum of Natural History, Smithsonian Institution, and Catherine Badgley, assistant research scientist at the Museum of Paleontology, University of Michigan. Dr. Behrensmeyer read the first draft, and her enthusiasm encouraged me to seek a publisher. Dr. Badgley reviewed the manuscript very carefully and critically, calling my attention to sentences whose meaning was unclear, providing correct terminology and usage when necessary, and suggesting stylistic improvements. Finally, and not least, I am indebted to Hans-Dieter Sues of the Department of Paleobiology, National Museum of Natural History, who also read the first draft and urged publication. He reviewed the final version and was helpful in providing correct and current geological nomenclature and the accurate spelling of German names. In addition, he took on the tedious job of editing the bibliography, supplying much missing information.

Publisher's Note

The University of Chicago Press joins Judith Schaefer in thanking Hans-Dieter Sues for his many contributions to the publication of this translation. In addition to providing many current taxonomic and geographic names, and reading the original manuscript and proofs, he has given a great deal of helpful advice and encouragement.

—

Preface

I recently had the opportunity to work for sixteen months on the Gulf Coast of the United States, stopping over at different places between the Mississippi River to the east and the Mexican border to the southwest. Although so many geologists who spend time in this area have complained about the lack of outcrops and thus the scant possibilities for doing scientific work, I found myself scarcely able, in the time I had, to study even incompletely the many questions of recent geology that presented themselves. The problems of sandbars, arid shores, subsidence, recent transgression, the contemporary upheaval axis, the *Taxodium* swamp, basins of brackish water, and the courses of rivers caught my attention, especially as they related to my research on the European littoral. This study of Recent carcasses and their paleobiological significance is a fruit of that sojourn, and I would be pleased if the reader who acquires information from the contents would also, as a natural scientist, make use of the methodology described in this book. Proceeding with a goal in mind can lead to success. In this kind of work we cannot always wait for a better or more impressive specimen to turn up; we must collect and compile. When we do come across a striking specimen, we replace a lesser one with it. I am aware that I cannot present a complete picture, but can only make use of characteristic examples. My focus is on the methodology of this problem, which has a paleobiological, a general paleontological, and a general geological component. The friendly reception given my remarks at the conference of the German Geological Society and the Paleontological Society at Göttingen tells me I am on the right track.

With a few exceptions, the photographs—of carcasses that I myself found and examined—are my own. To avoid distortion, I often took them with a hand-held camera looking down on the carcass from a point directly above it. It is my fervent wish that explorers and all those who live even temporarily in areas relevant to the subject matter think about what kind of observations of Recent geology we paleontologists need and understand how meticulous they must be if they are to be useful to us.

It is my pleasant duty to thank the publishers for their cooperation and, even more, for their understanding of the need for illustrations in a work such as this.

Some parts of this book will interest geologists and paleontologists, others climatologists and geographers, and still others biologists, perhaps doctors, and even hunters. I hope this book succeeds in stimulating interest wherever it can.

RECENT
Vertebrate
Carcasses

AND THEIR

Paleobiological
Implications

Introduction

Appreciation of the ontological method and its results has swung like a pendulum during the last ten years. At times, the method has been viewed with optimism and great approval; at other times, it has been met with pessimism and disapproval. In general, one can say that in spite of Walther and Abel, work done on Recent geology has for far too long been insufficient. Most of the papers are not systematic and goal oriented; the authors seem to be feeling their way along, analyzing isolated phenomena without attempting to integrate their findings into a more comprehensive point of view. It seems to me that the fundamental blame for this situation lies in the fact that even now people only occasionally try to apply basic geological inquiry to recent isolated observations whose causal relationship with more general phenomena has not been well explained. Thus, we often find well-studied fossil material being compared with Recent, insufficiently studied material.

To be more successful, a much larger share of the work done must deal with present-day phenomena than has been the case until now. What we see often seems to be so natural and obvious that we underestimate the importance of the problems presented. And this brings up another difficulty that causes much work to founder. Simple description of present-day phenomena is not geology, not by any means, and the findings of recent research should not be based on insufficiently researched and unexplained geologic phenomena.

Recently, studies have appeared attempting to make the results of hydrobiological research useful to geology; such an effort is very important to us geologists, but it seems that the authors are still groping their way along, sometimes getting bogged down on the border between the present and the geologic past. The setting forth of Recent relationships, in itself, does not contribute much to the science of geology; there is, however, a very valuable correlation, one that we ignore only at our own cost. The laws of the geological past throw light on recent finds, and the laws we establish in the present must be found applicable to fossil material. In the following treatise, I have tried not to sidestep this difficulty and have given the greatest possible consideration to the connection between the present and the past. The most important thing in such a study remains, however, the documented, accurate understanding of the specimen used for comparison.

The points of view and laws developed here do not have, of course, the validity of mathematical laws and should not be regarded only in a theoretical way. They are, in part, only probabilities, and we must always realize how easily the same results can arise from different causes. In such cases, insofar as it seems clear, we must have the courage to stand up for what we believe has been proven, or else, without coming to a conclusion, give the specimen up to someone else for further critical study. As far as method goes, we are in new territory, and there is still a long way to go before enough data have been gathered. The most important thing for explorers and others who spend any length of time in sparsely populated areas where fauna is abundant is to understand that it is not enough to know how many individuals died as a result of any given storm, drought, volcanic activity, or whatever; we need to know the manner of death, the position of the carcasses, the manner of their decomposition—this is the information we must have if we are to make general geologic observations about the embedding of vertebrate remains within the rock series.

Questions strike the visitor to the great collections: How did all these animals die? What happened to them before they were embedded? What particular conditions enabled their preservation in such great numbers? We shall try to give somewhat more detailed answers to these questions. The problem of how and under what conditions vertebrates died and became embedded has, for the most part, been skimmed over in the paleontological literature. It is the exception that detailed studies, such as Johannes Walther's research on Solnhofen, are dedicated to the answers to these questions. In my opinion, the reference works and textbooks on vertebrate paleontology close their chapters on general paleontology all too soon. An exception is Abel's *Paläobiologie der Wirbeltiere.* Wiman was completely right when he maintained that people were more interested in catastrophic mass death of animals before 1850 than after the disproof of the theory of catastrophism. It is time we deal thoroughly with the details of these geologic occurrences. The problem of diagenetic processes, which take place after embedding in stone, is also related to the issues discussed here. It is addressed in detail in Deecke's important book, *Die Fossilisation.*

1 Death and its aftermath

1. DEATH

In the dead organism, the metabolism characteristic of the living being is absent. Verworn's research has shown how difficult it is to ascertain the exact moment of the onset of death. Even when the heart has stopped, other organs remain active. As long as twenty hours after death, isolated hearts of children have been caused to beat again by artificially activating the circulation using Ringer's solution. Death is a process; indeed, the time of dying has also been given a name—*necrobiosis*. The duration of necrobiosis varies considerably: in a smaller organism it can take place within a few seconds, whereas in a larger organism an incurable illness can prolong it considerably. It consists of either the reduction of living matter by histolysis or the accumulation in the cells of substances foreign to normal life processes; thus, it is a metamorphic process.

Death from external causes can occur in various ways:

1. Withdrawal of oxygen from the surrounding medium results in death by suffocation for every aerobic organism.
2. Lack of food leads to death by starvation.
3. Deprivation of water changes osmotic pressure fatally.
4. Rising or falling temperatures cause severe hypnotic and static disturbances with fatal consequences.
5. There are many kinds of intrusive factors that can cause death.

They can take effect quickly, as when strangulation of the large carotid artery, which supplies the cerebrum with oxygenated blood, occurs, provoking complete unconsciousness after a few seconds, or slowly, as when humans suffer from chronic alcohol or nicotine poisoning.

Often an organism can overcome severe damage through its own self-regulating mechanisms. For example, persons who have drowned or been hanged have been revived hours later. Dormancy can readily lead to mass death if certain environmental conditions persist too long. So, too, can fish and frogs remain frozen for a long time and then revive—if the thaw is not too sudden. What happens is that the living cells adjust themselves to a bare minimum of metabolism, just as happens when animals hibernate. The important thing here is the complete independence of the nervous system from the cerebrum and heart. I have already mentioned that the heart has an enormous capacity for recovery. The lungs also play an important part, as is shown by resuscitation through artificial respiration.

Rigor mortis is a process that normally spreads slowly throughout the whole body and then just as slowly disappears. It can usually be noticed after ten hours, beginning in the head and spreading through the neck to the trunk. After an additional ten to eighteen hours, the condition passes. Then the skin begins to discolor, first where it is in contact with the surface on which the carcass is lying. With the lowering of internal pressure, the eyeballs shrivel and wrinkle, the cornea becomes cloudy, and the stench of death becomes apparent; putrefaction, beginning in the intestine, causes discoloration of the skin of the abdomen. Muscles of living organisms function as antagonists. After death, however, the flexors and extensors contract simultaneously so that all flexibility and movability become impossible. With increasing loss of heat from the warmer to the colder parts, the body contracts; but the warmth newly created by decomposition suspends this contraction. There are instances of rigor mortis setting in immediately, especially after great physical exertion, as has been noted in soldiers on the battlefield and in animals killed in bullfights.

As rigor mortis disappears, putrefaction and decay begin. The products of decomposition are simple substances such as carbon dioxide ("carbonic acid"), methane, water, hydrogen, ammonia, and nitric, sulfuric, and phosphoric acids. We all know of the great resistance of game to spoiling. There is a difference between spoiling and ripening, or autolysis, which has to do with a splitting of dead protein that is not provoked by bacteria. True "peak of flavor" is by no means decay or putrefaction, but a progressive degree of autolysis, or aging. Tainted autolysis (spoilage) occurs when decomposition is hastened by enzymes, causing an unpleasant-smelling substance to develop. It happens when the temperature is high or when the pieces of meat to be aged are piled on top of each other.

Decomposition of a carcass has two forms: one caused by aerobic bacteria, which attack the flesh from the outside, and the other by anaerobic bacteria, which penetrate the flesh through the wall of the intestine; the latter is the true cause of putrefaction of the carcass. Additional processes are set in motion by molds and by maggots hatching from eggs introduced by insects.

Discoloration of the dead body increases, and the activity of saprophytic kinds of bacteria causes the tissues to become soft and fluid, and they begin to exude foul-smelling gases. The ensuing swelling encourages attack by vultures and can cause the carcass to burst. Saprophytes obtain nourishment and energy by breaking down complex proteins. In this way, the organic components of the animal's body are used for respiration and fermentation. From the structure of the gases, one can see that certain saprophytes, active when no air is present, can break down proteins, which often contain 1.5 percent sulfur, and synthesize hydrogen sulfide. Hydrogen sulfide can also be formed when rotting organic substances come in contact with sulfates. Hydrogen forms during decomposition and captures oxygen from the surroundings. The presence of hydrogen sulfide–producing saprophytes can easily be proven by means of a test using lead paper, which quickly turns black if the gas is present. In contrast, it is striking that when a carcass is unearthed, a strong smell is never noticed. The reason is that the earth has the capacity to absorb most of the foul-smelling gases. This is also the reason that the shark described by Richter (134), which was buried in an oxygen-free soil, developed so little strong odor.

True aerobic decomposition does not begin until a certain degree of desiccation has taken place. After four years, all that is ususally left of a buried carcass is the skeleton and a brownish, humuslike substance—the remains of the rest of the body. Hampus von Post (Wiman 176), in his study entitled "Über die coprogenen Bildungen der Gegenwart" (1962), concluded that "the only thing left of the thousands of generations of animals and plants that have lived since the glacial era is a layer of soil about three to four centimeters thick." After the humuslike substance has formed, bacteria are replaced by molds, which can multiply profusely. For this process—so important for the inhabitability of the earth—to take place, aeration of the soil is necessary, and this condition is not always met at all places of burial. If the burial medium is water, it must remain fresh if the process is to take place.

If air is cut off, malignant putrefaction sets in, causing the formation of noxious substances. Under these conditions, adipocere is formed. This postmortem formation of adipocere as a process of decomposition is not an infrequent occurrence. The best known example is at the Cemetery of the Innocents, one of the oldest graveyards in Paris. Here, in 1786, during the installation of drains, about twenty thousand corpses were exhumed. In some of the coffins, which were stacked on top of each other, corpses were found whose soft parts had completely maintained their outward appearance. The same thing was observed in Zürich in 1912. A few years ago, when an old slate quarry near Probstzella [East Germany] was reopened, the carcass of a pig that had been thrown in was found. Its bones had been completely dissolved by water rich in alum due to the disintegration of the pyrite of the strata surrounding the seam of roof slate; the soft parts had been completely changed to adipocere. (The pit water of this eastern Thuringian roof slate quarry is well suited for conducting experiments to clarify questions raised in this regard.)

The curious process whereby fat, muscle, and bone become adipocere is a form of hydrolysis called saponification. It is based on the formation of fatty acid soaps from fatty acids such as stearine, oleine, and palmitin, and salts of ammonia, potash, or calcium. The yellowish, waxy substance thus formed is called adipocere, and it retains the fine details of the tissue it replaces. Even before saponification, there is usually a three- to four-week period of putrefaction, which then, however, takes place under the exclusion of air or with the help of antiseptic conditions in the ground. Fossil and subfossil hydrolysis of fish carcasses has often been described.[1] Abel (7) differentiates between the process of *Verwesung* [aerobic decay] and that of *Verfaulung* [anaerobic]; by the latter term he means the destruction of the organic substance in the absence of oxygen.

All putrefactive bacteria are sensitive to acid. Therefore, an acidic medium delays the onset of putrefaction, which cannot begin until the acids, mostly amide and amidic acid, are somehow neutralized. Putrefaction produces ammonia from dead proteins as well as from urea. Nitrobacteria can build soil nitrates, but putrefactive bacteria break down ammonia, freeing nitrogen into the air, exactly as in the breakdown of soil nitrate by denitrifying bacteria. Soil nitrates can also be formed inorganically by an electrical charge in the air above ground. In addition, organic phosphatids, present after death, are broken down by bacteria into insoluble soil phosphates. The fact that ordinary peaty soil is bacteria-free just below the surface is due to this acid reaction. Heath humus, peat, moor, and swamp soils are often rich in humic acid too.

In carnivores, digestion takes place in the large intestine. Chyme in the small intestine still shows an acidic reaction because of lactic acid fermentation of carbohydrates. Neutralization of the acid by the alkaline intestinal juices does not take place until the chyme reaches the lower part of the small intestine. Therefore, the chyme is neutral as it enters the large intestine and at that point usually contains one-seventh of its original available protein, which is then lost to the body through digestive putrefaction (see Omeliansky 123). Then the substances indole and skatole form (just as when the brain decomposes), giving the feces its repulsive odor; inside the body, they change to idoxyl and skatoxyl, combine with sulfuric acid and nucuronic acid and become harmless, just as are other poisons that

1. The first step in saponification, the breakdown of fats into fatty acids and glycerine, takes place in water. Even water alone, without the help of enzymes, can induce the process. The resulting product, adipocere, dissolves almost completely in alcohol, contains no glyceride, and is combustible. In an experiment, finely chopped mutton fat was placed in running water for two months; it changed almost completely into free fatty acids. Apparently, even at normal temperatures fat breaks down into glycerin and free fatty acids. Höfer sees in this process the first steps in the formation of natural crude oil. According to Deecke, the outline of vertebrate carcasses often preserved in the Posidonia Shale is attributable to previous saponification. Wiman observed whale flesh transformed into adipocere floating in the ocean, and a fish from the Bohemian Upper Cretaceous at Chotzen [ČSSR] was found preserved as an adipocerelike substance (see also Rüger, *Central-Bl. für Mineralogie etc.,* 1925). Formation of adipocere in human corpses is treated in a book by Müller: *Postmortale Dekomposition und Fettwachsbildung.*

form and then exist as hydroxyls. In this less harmful form, they can then be excreted with the urine. Decomposition outside a body gives rise to ptomaines such as cadaverine and putrescine; these substances cannot form within the intestine under normal conditions.

Because fossil coprolites are frequently found, the structure of excrement is of some interest to the paleontologist. According to Scheunert (140), feces are composed of the following substances: (1) metabolic end products and excreta from the mucous membrane of the intestine (mineral salt, intestinal mucus, shed epithelia); (2) products resulting from breakdown by digestive juices or their components (e.g, enzymes, hydrochloric acid, sterkobilin, coprosterin); (3) microorganisms; (4) indigestible components of food (e.g., insoluble salts, keratin); (5) digestible components of food that has been consumed in excess of need (e.g., elastin, cartilage, cellulose, and cell components, crude starch, pentosan); and, (6) unabsorbed products of digestion and decomposition (fatty acids, soaps, amino acids, hematin, phenol, indole). On the other hand, bones, fishbones, pieces of hide or skin, claws, and similar hard parts are not found in excrement as often as one might think, since sharks, as well as amphibians and reptiles, mammals and birds (pellets) are capable of regurgitating such things before they reach the intestine.

Human feces are made up, for the most part, of excreta from the intestine, materials of decomposition, and microorganisms. Only a cellulose-rich diet (e.g., mushrooms) produces an appreciable quantity of undigested foodstuff. The same is true for the feces of carnivores, while those of herbivores are mostly composed of undigested food.

The amount of feces resulting from different diets varies considerably. Sheep produce between 1 and 2 kilograms of feces per day; pigs, 2.5–3 kilograms per day; beef cattle fed a normal diet, 15–30 kilograms, and a fattening diet, 40–65 kilograms per day; horses on a pure hay diet, 16–17 kilograms, on a diet of oats, chaff, and hay, 9–10 kilograms per day. A meat-eating dog produces 27–40 grams; when fed bread 250–400 grams, and when fed large amounts of bread even 900 grams.

2. DECOMPOSITION

Herbivores such as cattle and horses can digest 30 percent to 75 percent of the cellulose content of their food, as can pigs, but humans can digest much less, and carnivores and birds none at all. We used to believe there were special digestive enzymes that could dissolve cellulose, but their existence has never been proven. Because the end products of cellulose digestion in animals are the same as in cellulose fermentation in sludge—acetic acid, isobutyric acid, carbonic acid, and methane—it must be a cellulose fermentation process. Because this process and its end products are so widely found in nature, they are evidently very important.

During the war, there was the problem of how to get the large numbers of corpses to decompose as quickly as possible so they would not become noxious. The efforts of the French were surprisingly successful. They hastened the process by applying the liquid that forms during the decomposition of cellulose. They conducted an experiment, using pig fetuses (to eliminate putrefaction caused by intestinal bacteria) weighing between 300 and 350 grams. The fetuses were either dipped into the liquid or sprayed with it. At a temperature of 30°–33° C, a sample taken after 108 hours was so completely decomposed that the bones lay in a reddish brown liquid. If the outside temperature was between 13° and 14° C, decomposition took 360 hours. The control samples, which had not been treated with the cellulose-dissolving material, were still intact after 66 days. Therefore, decomposition can be significantly accelerated, particularly at high temperatures, by addition of the enzymes that decompose cellulose. It was calculated that a human body would decay in 15 days using this method. A dog weighing 60 kilograms, placed inside a dung heap, decomposed right down to the skeleton in 8 days. In another experiment an artificial, sterile dung heap was built to avoid contamination and interference due to environment, temperature, humidity, and bacteria, and fresh fetuses were doused with water and urine. The thermometer inserted in the pile showed the onset of decomposition after 24 hours at a temperature of 16° C, compared to an outside temperature of 13.5° C; on the tenth day, the temperature had increased to 22° C, while the outside temperature was 16° C. From the eleventh day, the temperature stayed at 20° C for 3 days and then gradually dropped down to the normal level. On the nineteenth day, all that was left of the five 130-gram fetuses was a few bones. In 456 hours, then, 650 grams of carcass had been liquified. The drop in temperature between the fourteenth day and fifteenth day indicates decomposition must have ended after 336 to 360 hours.

We have already mentioned how widespread cellulose fermentation is in nature. We need only think of soil, with its vast quantities of plant remains, decomposition of which is revealed by swamp gas, which rises not only when the subsoil is disturbed, but frees itself through self-generated pressure. In an aquatic environment, the rising gas bubbles leave crater-shaped holes behind in the sludge. The formation of gas increases as barometric pressure falls. In late summer, all forms of life are flourishing; before winter begins, many die and sink to the bottom, where they are covered by sediments settling in the water. The formation of gas by decomposition does not subside until the cold has penetrated to the lower depths; the rising gas bubbles freeze separately, one above the other, forming a column within the ice cover. Gas is still actively forming at the beginning of winter, and the bubbles can grow to be one meter in diameter.

When animal carcasses are present during the decomposition of cellulose, as we shall see happened on the shore of Smithers Lake [Texas], the results are clear: Carcasses lying in water are completely reduced to skeletons in scarcely three months (see pl. 29, fig. A), while a considerable quantity of flesh and skin can still be found on individuals lying on dry ground (see pl. 31, figs. A–E). Animal remains can also undergo humification, just as plant remains do. The nitrogen content of the brownish product thus formed is always very high, and the product can be decomposed by mold and bacteria.

We have seen that when oxygen is lacking—in cemeteries that are too damp, for example—adipocere is formed. When it is desirable that corpses disappear as quickly as possible, well-ventilated burial sites are also unsuitable. In dry, sandy areas, we can observe that the soft parts shrivel quickly, and because of the dryness, the parchmentlike skin sticks to the bones, turning the corpse into a mummy. In many societies, the art of mummification has been developed from this natural process. When corpses are exhumed in central Asia, it is surprising how fresh they look, considering that they were buried many centuries ago. The "mummies" from the dry areas of North America and the west coast of South America are really such desiccated corpses, since embalming was not practiced by any American peoples. A number of European burial sites also exhibit this desiccation. The catacombs in Palermo [Italy], with their many mummified Capuchin monks, are famous. This kind of mummification is also found beneath the church on the Kreuzberg in Bonn [West Germany], in the castle chapel at Quedlinburg [East Germany], in the former Kaladulensier cloister on the Kahlenberg near Vienna, in the Franciscan cloister at Marientrost, near Graz [Austria], in the west crypt of the tower of Saint Michael in Bordeaux [France], and in the tombs of the Cordeliers and Jacobins in Toulouse [France]. There are also such corpses in the Green Vault in Dresden [East Germany] and in the lead cellar of the cathedral in Bremen [West Germany].

There exists an extensive literature on the mummifying properties in the Bremen cathedral. Although the real cause is a relatively cold, dry, continuous current of air, some have thought this mummification was due to radioactive properties of the lead cellar itself. Sander (136) concluded that there were quite a few underground vaults and caverns that had mummifying properties not fully explained. His explanation was that the radioactivity of the soil and of the cellar air had a direct effect on the organisms of decomposition and putrefaction and on the chemistry of fermentation. He even suggested that a throughly detailed examination of the relationship between radioactivity and fossilization should be made. To that end, Wigand (175) took measurements of the air in the Bremen cellar, and the results were completely negative. He could not find above-normal radioactivity in the air, in the walls, or in the lead coffins; radioactivity, therefore, was not responsible for the aseptic conditions of the lead cellar. The Breman cathedral is built on dry sand. The mummification of the corpses took place in the east crypt, and the mummies were not removed to the new lead cellar until they had already become desiccated.

Fermentation is stopped by both low temperatures and antiseptic substances. (The latter are effective partly because of their reactive properties, partly because they are used in heavy concentrations, and partly because they induce either dehydration or actual poisoning of schizomycetes.) We ourselves use these features in food preservation. Examples of mummification are really not so uncommon. The mummifying properties of the morgue at the hospice of Saint Bernard [Switzerland] were a consequence of natural ventilation by very cold, dry air; the same situation can prevent decomposition in polar areas and on mountain peaks, as is known from descriptions of Eskimo burials. On the other hand, the constant heat of desert sands can have a sterilizing effect, so that aerobic bacteria cannot do their work. Pausanias tells of the dried-up corpse of a warrior in the attic room of the temple of Isis at Elis [Greece]. Sander (136) tells of a French author who found in an airy, rain-tight place under the rafters of a family mausoleum the mummified corpse of a man who had hanged himself in a sitting position ten years before. Also not too uncommon are cases of natural antisepsis, which can often lead to mummification. In soil rich in saltpeter or iron, such as is found in moors, remarkable conditions of preservation can exist (think of the corpses of prehistoric men and animals found in swamps), and the shriveled but undecomposed corpses found in old salt mines figure prominently in tales and novels. Such corpses have also been found in copper mines. Concentrated hydrogen sulfide seems to be as strong an antiseptic as humic acid. In the Black Sea, the bottom water contains twenty times as much hydrogen sulfide as the water at 213 meters. Putrefying organic substances sink to the bottom and react with hydrogen sulfide from the existing sulfates. Sulfate-reducing schizomycetes play an especially important role in this process. If the body of water is still—special conditions of the salt content favor this—then hydrogen sulfide accumulates, and we admire its capacities for preservation in the Posidonia Shale and in many other localities.

The interesting specimen decribed by Richter (134) also belongs in this chapter. It is a shark carcass that lay almost intact on the Minsener Oldoog sand flat east of Wangeroog [West Germany] for one and three-quarters years. The sand is completely black because of its hydrogen sulfide content, and the carcass lay in groundwater. The preservational properties of hydrogen sulfide made it possible for the skin of saurians and sharks to be preserved at Holzmaden [West Germany]. Because fish in these strata were preserved in various positions, Hennig (68) once tried to ascertain the original shape of the body based on cross sections. He estimated that the compression in the Liassic Shale had reduced the body diameter to one-twentieth of its original size.

The embedding of animal carcasses in material that is long-lastingly antiseptic, such as mineral wax, asphalt, amber, and other resins, does not necessarily lead to mummification, as we see in Galicia [Poland] in carcasses of rhinoceroses and other animals in the ozokerite from Starunia

[Poland]. Even within the encasing medium, intestinal bacteria often cause liquifaction and gasification of the carcass. Phosphoric acid, too, seems to exert a preservational influence, displayed by the extraordinarily well-preserved remains of frog mummies and insects from Quercy [France].

Many examples of unusual or selective preservation of easily decomposed body parts can be found from the geologic past; a complete account of them is not possible here, but a few should be mentioned. Especially remarkable is the carbonized tassel from the tail of *Propalaeotherium hassiacum* Haupt (see Haupt 63), or the calcified tongue with brain attached from *Equus caballus,* from the Gyugy tuffaceous limestone, preserved in the Royal Hungarian Geological Institute (57). Well-preserved brains are not uncommon. Neumayer (119) described brain casts from silurid skulls, which were found with marine, brackish, and freshwater forms, and with leaf remains, in a deltalike deposit. In lower Miocene freshwater strata from Tombach, in the area around Eibiswald [Austria], Schmut (144) found a claylike structure the size of a walnut representing the brain cast of a turtle. Tilla Edinger (45) described the excellently preserved brains of *Diplobune bavarica* O. Fr. and *Anoplotherium* (the former found preserved along with its skull), found in the lower Miocene of Eselburg, near Ulm [West Germany] and now in the collection of the Museum of Natural History of Stuttgart [West Germany]. The results of Jaekel and Stensiö's research on the brains and nerves of the heads of Devonian fish are brilliant. Lull (103) described the steinkern of the cranial cavity of an odontocete found in a chalk bank in the upper part of the Lower Miocene. Andrews's find in the Gobi Desert—dinosaur eggs whose remarkable state of preservation I was able to admire in the museum in New York—attracted much attention.

The state of preservation of a bird egg found as detritus in the gravel of the Gila River in Arizona is also interesting. The egg, dating from at least as early as the Pleistocene (Morgan and Tallmon 117), was found inside a concretion and is filled with a yellow mass of crystalline colemanite. In some places near the shell, however, there is a dark brown, semiliquid, tarry material that resembles asphalt in appearance and physical characteristics. It is, in fact, natural asphalt, which has a conchoidal break with shiny surfaces when cold; as the temperature rises, it becomes softer, and at 100° C it liquifies with noticeable viscosity and a specific weight of one-third that of boiling water. It was precipitated before the deposition of the colemanite and, when heated, gives off many combustible gases so that it is then only partially soluble. This is an instance of direct transformation of the organic substance of the egg. Abel (7) also reported recently on the occurrence of fossil bird nests with the clutch preserved in the Tertiary sediments of western Nebraska. It is so well known that, in nature, cold is an excellent means of preservation that I need only mention the mammoth and rhinoceros carcasses that have come down to us. The extinct

Bison crassicornis [Bison priscus] found in the Alaskan tundra was almost fresh (see Lucas 101).

Observations such as those published by Wüst (184) are also very important geologically. He discovered remains of forage in the valleys between the tooth cusps of two rhinoceroses referable to *Dicerorhinus hemitoechus* Falc., found in 1893 in the Eemian strata in the Kaiser Wilhelm Canal [West Germany]. The purely vegetal material consisted of crushed plant remains, a little piece of resin, remains of leaves and bark, lots of dustlike matter, and thorns and woody remains from a rose, exactly as one finds in teeth in Recent skulls. Diatom-rich, noncalcareous sand stuck to the outside of one of the teeth.

3. PRESERVATION

Let us return to the fact that in nature, preservation by mummification is no oddity. When there is not enough air present, the reduction processes of putrefaction in the corpse are incomplete, leaving behind matter containing carbon and nitrogen, as in fractional distillation. On the other hand, when there is enough air, decompositional oxidation, comparable to a slow but complete burning, takes place, and almost all of the organic components are converted to gas, leaving nothing behind but the skeleton, which is insoluble in water and subject to oxidation only at very high temperatures. Mummification takes place when the carcass loses most of its water content before decomposition can begin its normal course. Albine's experiments in this regard are very important. Using a warm current of air, he expelled all the water from fresh rabbit carcasses, in an amount equal to 67 percent of the body weight. The artificial mummies thus created withstood putrefaction. Methods of salting and smoking are based on this extraction of water. Sun-dried fish preserved in salt were often found on the shores of the Caspian Sea and other bodies of water in arid lands, as Freyberg found on the west coast of South America (55), and as I found on the coast of the Gulf of Mexico.

The preservation of extremely shriveled yet quite well-preserved pieces of skin from the giant sloth, *Grypotherium domesticum [Mylodon domesticum],* in a cave in southern Patagonia [Argentina], is also attributable to mummification. Fonk observed near Lake Userekera, not far from Rufiji [Tanzania], the mummified remains of a rhinoceros that had not been a victim of human attack, but had been killed in a fight with several hippopotamuses, whose tracks had dried in the mud.

Mummies are easily formed after unnaturally high waters recede quickly. During a storm, when quicksand is formed even higher up on the beach than usual, animals sink in deeply, but as soon as the storm is over, air circulation is restored and the entombing sand dries out again (pl. 5, fig. E). Mummification can also occur during the unusually low water levels that occur during droughts. As rivers shrink and lakes recede drastically, the earth dries out as hard as stone. Swamps dry up too, and animals that have sunk into them and died from lack of food or other causes become desiccated or mummified, as I observed at Smithers Lake in Texas (pl. 31, figs. A–E). Also not infrequently, when water levels rise, completely mummified carcasses are once again in damp surroundings or areas of sedimentation. The opposite can also happen: partially macerated carcasses can dry out and become mummified. When desiccation is extreme, we often find the cervical spine curved sharply backward.

These complex events must be taken into account when we come across certain curved skeletons such as those found in the Kupferschiefer. Obviously, the unusual mummification of the *Trachodon [Anatosaurus]* from the Upper Cretaceous of Kansas has some connection with these processes; perhaps these animals died in quicksand that quickly dried out again, restoring air circulation, which caused the carcasses to mummify before decomposition could begin. During the preparation of the *Trachodon* mummy at the Frankfurt Museum [West Germany], the carcass of a fish was also found. In any event, the mummy had been embedded whole, and only later did the mummified skin and flesh decompose; what we see today is a pseudomorph of the originally mummified part.

4. THE ROLE OF INSECTS

When a human corpse, decomposed to the point that only the skeleton and a brown, humuslike substance remain, is exhumed, in addition to a covering of mold we usually find maggots and pupal cases of flies and beetles. If aeration of the soil is relatively good, their share in the process of decomposition is often considerable.[2] There is a simple way to mummify a bird carcass. If it is wrapped in paper so well that flies cannot lay their eggs on it and stored in an airy place, the result will be a wizened, odorless little mummy. The absence of flies is important; their role in decomposition is to begin the liquefaction of the body. They lay their eggs anywhere the hatching larvae can readily penetrate the carcass—on the bill, in nasal openings, in the eyes, in the anus, or in wounds. (One can refer to medical books to see how frightfully living human beings in tropical lands can be ravaged by fly larvae. For example, I think of a photograph from the Hocrouz Institute in Rio de Janeiro [Brazil] that shows a patient with over half of the forehead and temple, and one eye socket, eaten right down to the bone by maggots.) Even cold temperatures, which, by the way, do not kill putrefying bacteria, but only cause them to

2. Evidently, the corpse of a nineteen-year-old youth, which had lain not longer then twelve days in the Grünewald, albeit in hot summer weather, was entirely skeletonized by fly maggots. When temperatures are favorable, maggots hatch quickly and begin the liquefaction and consumption of the carcass. The considerable capacity of ants to skeletonize is common knowledge.

go dormant, cannot harm fly eggs. When the snow melts, putrefaction and hatching of fly eggs resume immediately.

Maggots leave a noxious carcass before pupating not only because it is a transitory environment, but for another important reason: only they—and not the egg or the pupa—can survive in an anaerobic environment. We see this most clearly in *Psilopa petrolei* of California, a fly only two millimeters long, whose larval stage is adapted exclusively to crude oil. These water flies inhabit asphalt pools in California and deposit their eggs right at the edges of the pools. The maggots hatch, enter the pool, swim near the surface, and feed on dead creatures floating there. The maggots of another form of ephydrid fly live exclusively in urine; those of yet another, only in saltwater; and the most well known are those that gather to pupate on the banks and beaches of the Great Salt Lake, in Utah. The Indians used them for food, making a broth and a kind of bread out of them. The emergence of maggots ready to pupate is used as a simple method of feeding fish in a pond: Wire baskets containing meat scraps and carcasses are suspended above the water by mounting them on stakes. The flies go about their business, and eventually a continous stream of fat maggots falls to the surface of the water to feed the fish.

Pupae also exert enormous efforts to burrow to the surface from deeply buried chrysalises. Fabre (46) has done detailed research on this subject. He placed 15 fly pupae in each of several test tubes and filled each with a different quantity of fine, dry sand. In one test tube filled with 6 centimeters of sand, 14 completely developed flies reached the surface successfully; in one filled with 14 centimeters of sand, 4 succeeded; and in one with 20 centimeters of sand, only 2 succeeded—the rest died trying. When covered with garden soil to a depth of 6 centimeters, 8 flies out of 15 pupae reached the surface; when covered to a depth of 20 centimeters, only 1 was successful.

Competition for spoils among the carrion-eating insects, many of which are beetles, is intense and causes, for example, the family known as sexton beetles to bury what they have found. The botanist Gleditsch performed an experiment with four of these. In fifty days, the beetles buried two moles, four frogs, three small birds, two grasshoppers, the entrails of a fish, and two pieces of cowhide. Insects are by no means rare in the phosphorites of Quercy (Gaillard 58). Thévenin (162) expressed his surprise that this insect fauna lived in the dark, in carcasses. All dipterids live in carcasses, and it is their larvae, which resemble the maggot of the modern warble fly, that occur here in such unusually large numbers. In addition, pupae from moths, a few large beetles, and some grasshoppers and centipedes were found. As in the "frog mummies" of Quercy, the insects themselves have almost disappeared, and the cavity left behind has been filled completely with phosphate.

Wanless (Abel 7) describes a rhinoceros skull from the oreodont beds in Nebraska that shows numerous bore holes made by carrion-eating insects; they prove that the bone lay in a certain position before being embedded. This phenomenon should not be confused with holes bored subsequent to burial, as happened to exposed skulls in the loess walls of Thuringia.

5. What Happens to Carcasses Lying on the Surface of the Ground

Over and over, the question arises: Where are the animals that die a natural death? The fact that we come across dead animals relatively seldom has given rise to many tales, especially in the case of elephants, whose valuable tusks cause people to devise preposterous schemes to search for their "dying grounds" or "graveyards." Indian tales, in particular, go into detail on the subject of dead animals that disappear in unnatural ways; it may even be that the belief in transmigration of souls has been influenced by this seemingly inexplicable phenomenon. Much that has been written on elephant carcasses refers to the descriptions by the elephant hunter Sanderson, who, even though he wandered throughout British India for several decades, came across elephants that had died a natural death only twice. Whether the rumors of the legendary, often-sought African elephant graveyards can be traced back to the Indian fables or whether they are indigenous to Africa is hard to say. Only a few years ago, a newspaper reported that the duke of Orleans had outfitted an expedition "to look for the exceptionally rich African elephant graveyard known only from tales told by the natives." The article was accompanied by an outlandish illustration entitled "The Duke of Orleans Hunts the Weird and Secret Billion Dollar Ivory Graveyard to Which the Great Beasts Creep Pathetically When Death Calls." This fabulous elephant graveyard is described as an inexhaustable source of ivory, potentially more profitable than any gold mine. Even today, ivory is worth ten times more than silver. Such a report only emphasizes how remarkably seldom travelers have seen dead elephants. When one finally does come across a carcass, death has been either violent or accidental. And so the legend that elephants have a secret dying ground to which they withdraw when the hour of death approaches comes to mind. Some natives even believe that the animals bury their dead. When the European protests that elephant graveyards could not exist, the native always responds emphatically, "Have you ever seen a dead elephant? And have you ever heard of anyone who has found one?"

Surveyors with decades of experience who have gone deep into elephant country have often expressed their surprise that although they have come upon several thousand living elephants in every possible place, they have never seen a single skeleton of a dead animal, except of one that had been shot. Most stories of animal graveyards are probably based on mass deaths caused by adverse conditions. There is said to be an elephant graveyard in India, east of Adams Peakros, accessible only by way of a narrow, steep-walled pass, on the other side of which is a clear lake. Supposedly the animals lie down next to it to die. The body of water in this tale is important. Sick animals suffer from thirst and must seek water. When they find it, they may just sink into the mud or perhaps, once they have entered the water, they never leave it again; then they are gnawed on by crocodiles or carried away by the next high water, just as the African traveler Schomburgk reports.

According to the tale told by the natives of Somaliland [East Africa], the elephant graveyard there is unusually large. Some people believe African elephants can cover long distances even when sick. The many elephants that used to live on the slopes of Kilimanjaro, Kenya, and Ruwenzori, and on the banks of the Victoria-Nyanza were supposed to have gone to this dying place—a deep valley surrounded by thousands of square miles of almost impenetrable forest in the interior of Somaliland. Reports from South America and Antarctica would have us believe that different kinds of seals also have preferred dying places.

All these tales have their origin in completely natural occurrences. Carcasses bloated taut in the heat collapse quickly, leaving the remains hugging the ground (see p. 1, fig. B). In a short time, they are reduced to skeletons by insect larvae and scavengers large and small. Other scavengers bury what is left, and in such a well-fertilized spot, abundant vegetation takes over if it is not too dry. I have only to think of the enormous carcass assemblage at Smithers Lake, to be described later, which just ten months after the catastrophe was covered with vegetation five meters high; grass covered the ground so thickly that only here and there did a protruding turtle shell or fish bone give scanty evidence of the considerable mass of bones concentrated beneath.

Without a doubt, animal graveyards are mistakenly considered to be places consciously sought out by dying animals, when in reality they are places where an entire herd died from a common cause. This is perhaps what happened on the scrubby banks of the Santa Cruz [Argentina], where the earth is said to be white with guanaco bones. These concentrations are most conspicuous when annihilation is due to an epidemic such as hoof-and-mouth disease. This disease is by no means limited to domestic animals, as its transmission to the deer and European bison of the Bialowicze Forest [Poland] proves. Moreover, the enormous devastation caused by African rinderpest, in 1890, affected not only the herds of the Masai, but also buffalo and other wild animals. After the epidemic of goat mange in the Alps, skeletons of animals that had died of the disease were often found.

On the coastal prairie along the Gulf of Mexico, one always sees dead cattle that are not among those killed in great numbers by winter storms or epidemics. Feeding conditions during droughts are often so poor that individuals sicken, perhaps contracting kidney inflammations from drinking bad, salty water or eating dry, indigestible forage. The slowly dying animals almost always fall to one side with legs and neck outstretched. In a short time the carcass swells enormously due to intestinal gas (pl. 1, fig. A), causing the hind leg lying uppermost during rigor mortis to rise to the horizontal or even higher. In extreme cases, gas pressure can force the carcass to roll onto its back or even completely over. The foreleg is only slightly raised because the area around the lungs and pectoral girdle deflates more quickly than the abdominal area. The anus protrudes considerably, facilitating the efforts of vultures, which always attack there first (pl. 2, fig. D). The vultures' feeding quickly releases the gas pressure and the whole thing collapses, the jutting hind leg falling usually a little in front of the one lying beneath it (pl. 1, fig. B).

It is astonishing how the carcasses, so enormous and conspicuous when bloated, seem almost to melt into the ground. This was also observed many years ago in Texas by Keilhack. Under the burning sun, part of the flesh dries onto the bones and does not decompose until much later. How much is eaten by scavengers depends almost entirely on how much food is available to them. Plate 2, figure D also shows that the legs have been carried off by scavengers. The scapula and forelegs are especially easy to remove, the hind legs much more difficult. Plate 1, figure C shows forelegs that have been carried off. The bones are still held together rather well by the tendons. In one, the hoof and shoulder blade are almost touching, and the bones of the upper and lower leg form a straight line, whereas in the other, the bone of the lower leg forms almost a right angle with that of the upper, which in turn forms a straight line with the shoulder blade. We can compare Osborn's illustration (124, p. 146) with these pictures; he shows a sketch of the remains of skeletons of six Pleistocene horses embedded in sand at Rock Creek, in Texas.

When the carcass disintegrates, gravity comes into play, especially when the underlying surface is uneven. Thus, we see in plate 16, figure A, the carcass of a horse (sometimes these are skeletonized by vultures in one day) in an unusual position. Although the flesh has already been removed from the skeleton, the ligaments still hold it completely together. The legs are stretched out toward the rear. The pelvis lies with its dorsal side up, and the rib cage is somewhat crooked because the skeleton is supporting itself on the spinal processes and ribs of half of the body—a frequent reason for an asymmetrical dorsal position. The ribs, no longer connected, tend to separate and are pulled partly toward the pelvis and partly toward the neck. Influenced by the flat bank, the neck is dorsal as is the skull (unfortunately obscured by a bush), whose lower jaw has been carried off. I took the photograph April 19, 1925, at a stagnant lake near the Brazos River [Texas] between Rosenberg and Richmond, a little south of Fulshear. Plate 16, figure B also belongs here; it shows the carcass of a cow on the edge of a beach cliff on Matagorda Bay [Texas]. Half of it has been washed away, and the carcass, pulled by gravity, hangs over the edge. We see here how quickly the destruction of such cliffs progresses—they recede even before the carcass decomposes. When sand is blown away by the wind, remains of vertebrates and excrement, under the influence of gravity, are often concentrated in the deepest hollows of dunes or projected into wind-eroded cavities—places where other material gleaned by the wind collects. When remains are covered quickly by windblown sand, there is a chance that they will be preserved.

In illustrated newspapers, one occasionally finds pictures of the widely scattered remains of pack animals that died crossing the North African dessert. Recently, in an American newspaper (60), there was a picture of a very typical corpse of a human who had died of thirst in the desert. The bones on the upper side were clean, but on the underside the partially mummified flesh was preserved. The arms were spread apart, and so, to a lesser extent, were the legs; the toes, however, all pointed in the same direction, as if the corpse were walking. Most deaths of this kind are the consequence of underestimating the amount of water needed for a journey across the desert.

No basic work in forensic medicine bearing on the question of how long a skeleton may have lain in the earth has yet been done. Exhumations from the Stone and Bronze ages teach us that the period of time can be considerable. One is therefore advised to make estimates with care. The degree of brittleness of the skeleton gives no information as to age; the nature of the place where it is found must be taken into account. The more the bones are exposed to air, the more quickly they disintegrate. The quantity of precipitation, the number of days below freezing, covering with clay, burial in sand or loam—all these factors play an important role in forensic medicine.

In 1925, in the Mojave Desert, the corpse of a gold prospector was found, probably dating back from the first California gold rush. His death evidently followed a strenuous fight with Indians, because gold nuggets were found strewn on the ground around his bones, which even after decades were still lying intact on the surface.

Although it is generally true that in the forest, decomposed game is quickly skeletonized, the bones disappearing within a few years, there are exceptions. In March 1925, the American hunting magazine *Field and Stream* published an extremely interesting picture taken in the Flathead National Forest, in northern Montana; it shows the carcass of an enormous elk and the body of the hunter who died during a struggle with it. The old-fashioned weapon indicates that both skeletons lay on the surface of the ground for several decades, without the elk's rib cage having collapsed. Even the cervical vertebrae were still connected. Only the head, weighed down by the antlers, had become disarticulated, and it was resting on its base; the lower jaw lay on its side next to it. The remains of the human skeleton lay a short distance away, along with the elk's foreleg.

The ability of birds to skeletonize is extraordinary and is not limited to vultures or birds from exotic lands. It is astonishing how thoroughly a titmouse can skeletonize a carcass that has been skinned and hung up. I remember the carcass of a fox left hanging in a tree; all that remained was a completely articulated skeleton. C.G. Schillings has described in detail how all the thousands of carcasses of animals that die on the open game preserves on the African savannahs disappear. The most striking example is the rapid decomposition and dispersal of the enormous pachyderm carcasses. Disintegration of the bones, mechanically and chemically induced, proceeds more slowly and sometimes does not happen until many decades have passed; it occurs when the phosphoric acid of the bones is displaced by the nitric acid of the soil. Many bones are also gnawed and eaten up; we often see bones in the forest bearing the marks of rodent teeth. In Texas, one often sees even cattle gnawing on the bones of their fellows who have died during the winter. Lichens grow, insects burrow, moisture softens, and heat cracks. Abel (7), too, has recently reported on remains that have lain at least three or four decades on the surface of the ground, exposed or slightly covered with windblown material. He brought back from Nebraska for his collection in Vienna bison skulls and skeletal parts that are still reasonably well preserved (see his fig. 178, p. 280). On the other hand, he shows another skull completely riddled by hailstones.

In Texas, certain ticks transmit a devastating rinderpest to livestock. To prevent the ravages of this disease, large Indian cattle called zebus were used to strengthen the bloodline. These animals are considerably more resistent or even immune to ticks, and, during the last ten years, interbreeding has brought about radical changes in the appearance of the cattle on the Gulf Coast. In many regions, the tsetse fly prevents keeping certain domestic animals, and malaria, carried by mosquitoes, is one of the main reasons for the use of Negro labor in those parts of North America where cotton is grown. When many animals have died from drought and cold winter storms, the carcasses lying around often cause the spread of epidemics, which can almost depopulate wide areas. Texas was infected relatively late by hoof-and-mouth disease—later, anyway, than Maryland and California. Often the centers of epidemics are found near harbors. After an incurable sickness broke out among muskrats in Bohemia [Czechoslovakia], people thought for a while that they were rid of the plague of these harmful rodents. But the resulting mass death by no means fulfilled their hopes. An epidemic seldom, if ever, causes the extinction of a species; it may, however, bring about vast fields of carcasses.

Not only vultures and scavenging gulls, foxes, dogs, wolves, hyenas, and other mammals, but crayfish, shore crabs, and many fish, especially eels, are excellent skeletonizers. The achievements of piranha fish are legendary: despite their small size they can attack any living animal with their ferocious jaws and quickly reduce it to a skeleton. The movie of the Dungern expedition showed how in only fifteen minutes these insatiable marauders of Brazilian rivers completely removed the flesh from an adult capybara, leaving only the cleaned skeleton. Predators and scavengers contribute to the preservation of remains when they bury carcasses.

6. THE EMBEDDING MEDIA

The dangers threatening vertebrates are mainly natural events, which do not always kill directly but often weaken and exhaust animals so completely that they eventually die. Some examples are earthquakes; volcanic activity with ashfall, mud flows, and gases; dust, snow, and hail storms; lightning; tornadoes; plunging temperatures; heavy precipitation; freezing rain; dry spells and drought; a winter with heavy snowfall that lasts too long; infected sources of water and salt; ticks and flies as carriers of epidemics and causers of panic; lack of substances necessary for life; high tides and inrush of water; flooding and high water; a change in the salt or gas content of water; a change in the substrate (as in the sinking of the land); sudden interference in the biotic community; gradual worsening of the climate; and on and on.

Numerous dangerous moments arise due to animals' instincts for self-preservation, manifested in the following: migration, during which animals encounter countless difficulties; reproduction, which often causes animals to wander away from their accustomed surroundings; hunger, which leads them to unfamiliar ground; thirst, which leads animals to water holes where carnivorous mammals and crocodiles lie in wait and which also causes sick animals to become stuck in the mud; the herd instinct, which often leads to destruction; panic and fear, which cause herds to stampede; again hunger, which drives inexperienced, young animals and weakened old ones into mortal danger; the need for warmth, which often causes animals to go into water or caves or forces them to crowd together out of the wind; the impossibility of avoiding enemies; and degeneration through inbreeding. Many different kinds of accidents can happen, too: falling while running away; breaking through ice; falling with snow cornices or being carried off by ice flows; burial by an avalanche; drowning; suffocation; being stuck in mud, swamp, bog, tidal flat, or asphalt, and so on. Isolated accidents can take place, but tend to happen repeatedly in places especially conducive to them, so concentrations build up that resemble those resulting from mass death caused by epidemic, famine, earthquake, and the like.

The actual cause of death is often a complex matter. We need think only of people who starved to death during the crisis of wartime; they did not really die from hunger but rather from some kind of acute disease afflicting an already weakened organ. We should distinguish between the internal, actual cause of death and the external circumstances surrounding it. Thus, famine caused by drought is unusually harmful to ruminants if they are then subjected to the drop in temperature accompanying a severe norther; even when they are so chilled, they do not simply freeze to death, but rather, in most cases, contract a kidney infection because of ice buildup on their hides. The urine becomes bloody and the overtaxed animal dies.

Exhaustion is an unusually common cause of death, as can be observed in migratory birds, in herds that have encountered a storm, in animals hunted down by predators, and in weak individuals that break down when a herd stampedes. Klähn (88) has made a sharp distinction between the two main groups of causes of death: dying and being killed. By *dying* he means normal death due to old age or sickness. By *being killed* he refers to vigorous individuals that become victims of accident, enemies, or the forces of nature.

Those that simply die are, for the most part, not preserved, because they succumb to sickness or normal old age in places where no sedimentation takes place, while an animal that is killed is usually linked to a place where it is likely to be embedded in the geologic record. Klähn takes this point of view: "Almost all mammal remains are from animals that died unnatural deaths." This coincides with Wiman's view (176) and with my own. Klähn supports his view by pointing to the degree of wear on the teeth, which, in fossil animals, has almost never advanced to a point that would indicate the probability of death due to old age. The forces of nature often kill herbivores and predators together.

Although Klähn has already admitted that here and there accidents sometimes happen in ways that do not involve pursuing predators or the forces of nature, Dietrich (41) has recently objected vigorously to the attribution of every collection of bones to accident or catastrophe, to violent death or natural events. This reaction is completely understandable, and the truth probably lies somewhere in between. However, the mass of evidence from strata really rich in vertebrate remains discloses overwhelmingly often the effects of single or repeated catastrophes, enabling us to say with some assurance that the vertebrate paleontologist deals mainly with animals that have died a violent death—victims of accident, thirst, epidemic, and climatic extremes.

In every landscape there are features that can become traps for animals. Individual animals are seldom victims of one of these; what usually happens is that during the migrations and movements of herds, individuals become careless, find themselves in damp places (or even seek them out to cool their feet), fall asleep, and then cannot get out; or perhaps weak animals drown; others fall off cliffs. Young animals are especially vulnerable because their performance under stress is inadequate. Usually such animal traps are more suited to sedimentation than the surrounding area. Carcasses attract scavengers, and there are animals that lie in wait for the scavengers too; one thing easily leads to another, and remains accumulate. In the Eocene of the Geiseltal [East Germany], the vertebrates described by Barnes (19) died and were buried in an old, swampy water hole that was probably once an abandoned river channel. It is possible that spring water rising from a sink hole kept this water hole open. Such natural animal traps and storehouses can be formed in many different ways.

The geographic configuration of landscapes makes it unavoidable that, because of complex conditions, places that store and protect are found in the middle of erosional regions. Such natural receptacles often lie on the very border between erosional and depositional areas or where land and water meet. Thanks to the ways that shorelines shift, there are coastlines where carcasses of larger vertebrates are washed up, usually at indented alluvial shores between points where destruction takes place. These wind-drift shores often show heavy concentrations of remains. The fossil whale cemetery of Antwerp [Belgium] and the recent one on the Bay of Biscay [Spain and France] are examples of this type of phenomenon. According to Johannes Walther (170), "at Cap Bessel, near Spitzbergen, the ground resembles a softened clay threshing floor, in which little trickles of water coming from the snow-covered slopes have built a network of trenches. What struck the eye was the masses of whale jaws, which lay partly on dry land and partly buried deeply in the mud." We hear repeatedly of entire pods of whales being stranded. In 1925, seventy-five black whales five to six meters long were stranded along the coast of Massachusetts. The *Berliner Illustrierte Zeitung* of August 30, 1925 (34th year, no. 35, p. 1119) carried a picture of this very characteristic festoon of carcasses (pl. 34, fig. A). The pilot whales of the school were mostly full grown and had run ashore along the coast of East Brewster, near Cape Cod. In this case, the herd instinct became the animals' undoing. They followed the old male leaders even when, confused, they turned toward shore. Minimal development of the labyrinth of the inner ear makes orientation difficult for these whales. Similarly, sometimes during flood tides whole herds of land vertebrates may plunge into the sea. Plate 2, figure A shows a large dead whale that had originally drifted ashore at the mouth of the Sabine River [Texas]; it was later carried off parallel to the shore between the mouths of the Sabine and Calcasieu rivers to a spot where alluvial deposits can be seen. In earlier years, I observed the washing-up of carcasses of harbor seals and common porpoises along the slightly indented coastline halfway between the Gohrener south beach and Lobber Ort, on the Isle of Rügen [East Germany]. The directions of currents and shifting coastlines destine certain places to be the sites of such concentrations of remains.

Water holes and seeps of tarry crude oil are animal traps of the first order. Nesting sites of birds are often used for many generations, and when they belong to predatory birds, considerable quantities of bones accumulate. During flooding, parts of the landscape that remain high and dry like islands become refuges for surviving animals; they crowd together in these places, but often succumb anyway to high water or starvation. Areas of bare rock serve the same purpose during grass and prairie fires.

Dune valleys in dry areas concentrate remains, especially when they surround temporary watering places. Floods create fringelike carcass assemblages on the periphery of high watermarks. Drought confines living things to a restricted area. Animals break through the hard surface of salt pans to the underlying hygroscopically damp clay, and areas of quicksand claim their victims. Obstacles encountered during migration increase the number of carcasses. Snow and dust storms force animals to crowd together in relatively protected places and decimate the inhabitants over a wide area. Remains at feeding grounds of crocodiles and warm-blooded carnivores have come down to us fossilized, as have deposits left by scavengers and nests of reptiles and birds, such as the ostrich nests in the loess in China and the dinosaur eggs in the Gobi Desert. The massing together of fish, amphibians, and reptiles (crocodiles and turtles) for reproduction is important and in many kinds of fish is followed by physiologically normal mass death. Carcass assemblages are caused by a change in salinity, gas content, and temperature of bodies of water. Cold waves and dropping temperatures claim their victims, as do arid areas, which are often afflicted by drought. Shifts in the courses of rivers also cause vertebrate carcasses to accumulate; so do volcanic ash falls and suffocating gases. Some vertebrate localities of the American Tertiary may have formed when the temperature dropped suddenly at the onset of a cold wave.

Caves in the most diverse areas are very effective storehouses for the remains of mammals, including humans. Often the animal contents are an important indicator of change in the landscape and climate of the surrounding territory. For example, according to Obrutschew (122), the fauna of the cave at Balagansk [USSR]—bear, deer, and roe deer—is very young, but gives evidence of earlier forestation of this area, which today is completely steppe. The reverse is true in Siberia; today forest covers a wide area, but in the time of the loess formation, mammoth and rhinoceros, bison and wild cattle, elk and caribou tried to take shelter in protected places during snow and dust storms.

It is impossible to list all the causes of death here. In certain instances, death results from the interplay of a number of potential causes.

Sometimes we can determine the cause of a mass death in the animal world indirectly from the composition of or change in the burial medium. Osborn (124), more than anyone else, has completely understood the enormous importance of studying the embedding media in order to form an opinion as to the origin of vertebrate strata. We have him to thank for the following summary of the main sediments in which mammal remains are found.

1. Conglomerates—with water-tumbled pebbles, gravel, and sand, characteristic of advancing and retreating coastlines, of mountain streams that spread out over the plains, or of riverbeds. "Mudball

conglomerates" are found fairly often in the Tertiary of western North America, and clay balls indicate just as often the location of reptiles in Europe. In the extensive Oligocene Badlands, there are coarse riverbed sediments overlying fine flood-born sediments, and each has its characteristic mammalian fauna. Not uncommon in these sediments are fossils transported by water to secondary burials.

2. Sandstones—often composed of quartz, sometimes also of feldspar or other grains. Sea and river sands cover larger areas than do the larger conglomerates with which they are associated. Sandstones owe their origin to slower movement of water or to intense wind activity.

3. Shales—mostly fine mud sediments built up in calm or relatively calm water, exhibiting a more or less completely horizontal splitting of the layers. The sediments often contain splendidly preserved leaves and remains of fossil fish, as in the famous Green River Shale; however, they seldom contain mammal remains and only occasionally bird remains.

4. Clays—built up in the floodplains of rivers and in deeper waters of oceans, wherever curents permit formation of uniformly fine sediments. Often the clay is a consolidated, loesslike substance. True plastic clay is mainly of marine origin. The famous London Clay, which yields *Hyracotherium,* is an estuarine formation.

5. Loess—characteristic of the Late Pleistocene. It is an unconsolidated, porous, gravelly silt, resembling certain sediments from areas subjected to flooding, and is spread by wind over dry land. In some areas it is composed of modified volcanic ash. The pampas loess and the Chinese loess are especially rich in mammals; the steppes of southern Russia are also significantly rich in bones.

6. Volcanic ash and tuff—typical burial media of the mountain basin formations of North America. Wind-driven ash can become very similar to loess.

7. Lignite—brown coal.

8. Gypsum—develops during evaporation of lagoons; always an indicator of climate. A typical example is the upper Eocene of Montmartre [Paris Basin, France].

9. Limestones—often of organic origin and mechanically deposited, as in the coarse chalk of the upper Eocene.

10. Marls—loose, unconsolidated deposits of clay, sand, and shell; often rich in organic matter, sometimes yields phosphate.

11. Phosphates—often associated with bone concentrations or excrement.

12. Asphalt—the residue of pitch lakes formed after evaporation and oxydation of crude oil springs; characteristic of the Pleistocene of southern California.

13. Breccias—can be either intercalated or superimposed.

Osborn, however, goes one step further. He maintains that organic inclusions and the composition of sediments permit us to clarify the geography of the prehistoric landscape. In so doing we must realize, of course, that we can only study the area of deposition, and that we must arrive at a conclusion as to the character of the erosional region indirectly, something Wepfer (173) considers in particular. Thus, the embedding rock and the composition of the characteristic beds are almost as important as the animal remains themselves. Osborn groups the most important paleographic units in which fossil vertebrates are found as follows:

1. Marine deposits—sediments built up in the ocean or at the edges of oceans, which occasionally contain remains of land and freshwater animals mixed with marine mammals and mollusks.
2. Estuarine deposits—brackish water deposits along entrances to mouths of rivers, characterized mostly by vertebrate remains.
3. Fluvial deposits—freshwater sediments in river channels, in ox-bows of stagnant water, at mouths of streams, or in deposits formed by torrents.
4. Lacustrine formations—freshwater deposits transported to lakes by rivers or streams, which, beyond the coarse-grained drainage areas, can be of the finest grain size and layered through periodic sedimentation.
5. Flood plain formations—deposits formed by flooding of exposed bottom lands. Periodic flooding can also cause extensively layered deposits.
6. Lagoonal deposits—in old river basins and bays, gypsum and other salts can precipitate due to evaporation.
7. Aeolian deposits—continental deposits of usually fine, unlayered sands, rock flour, and dusts, always lacking the horizontal bedding of limnetic and flood deposits. Such rocks are characteristic of the North American Late Tertiary.
8. Cave deposits—the formations are composed mostly of weathered residue or masses of alluvium. Often found are loams and deposits laid down chemically by trickling water.
9. Fissure deposits—heaps of bones transported and deposited by wind and water, and more or less consolidated; characteristic examples are found at Quercy [France] and Egerkingen [Switzerland].

[This paraphrase from Osborn (124) has been translated from the German to conform to the original text—TRANS.]

2 Modes of death

1. DEATH DUE TO VOLCANIC ACTIVITY

A large number of vertebrate localities testify to the disastrous effects of volcanic activity on the animal world. There is even geological evidence of animals fleeing volcanic activity in panic. Thomas von Szontagh, former director of the Bureau of Geology in Budapest, found such evidence in the form of tracks on Miocene sandstone slabs in the district of Ipolytarnoc, Nograd Komitat [Hungary]. Volcanic ash covers rhinoceros, deer, bird, and other animal tracks imprinted in the stone. I saw a large, interesting slab from this stratum in the museum in Budapest. Even more well known, from historic times and of verifiable date, are the corpses excavated at Pompeii and Herculaneum [Italy]. Bodies of humans and animals are found in exceedingly natural positions, just as they were when death—from poisonous gases, ashes, falling debris, and so on—surprised them. These corpses were not excavated in the true sense; instead, casts were made of the cavities left in the fine ash by their bodies, long since completely vanished. Oddly enough, we have so far not come across similar cavities made by carcasses in ashes and tuffs of earlier geologic times; only cavities corresponding to tree trunks are known.[3] We find the corpses at Pompeii in very typical dorsal and ventral positions; the knees of those in the dorsal position are usually partially flexed.

It would be beyond the scope of this work to explore in detail the relationship between volcanic deposits and concentrations of vertebrate remains, but the recent geologic observation made by Johannes Walther on this subject (171, p. 826) in the area of volcanic activity in the Dutch East Indies [Indonesia] are important:

> This type of volcano has crater lakes and mud basins: during massive eruptions, cracks form in the walls around the lakes, and the masses of mud found on their bottoms flow downhill. When there is

3. As a geologic analog, Deecke puts forth the elephant carcasses embedded in tuff and the tree trunks from buried forests in the Roman Campagna.

heavy precipitation at the same time, and the lakes rise, these "mudlavas" devastate the area for miles around. On October 8, 1822, a mudflow mixed with hot water and rock rubble poured out of Gelungung Peak and covered all the villages in the vicinity, turning the area into a steaming, blue-gray cesspool, thick with corpses of men and animals, rubble from houses, and broken tree trunks. The streams Tji-Kunir and Tji-Wulan broke unchecked through these masses of mud and rubble, swelled to raging torrents, and destroyed everything in their paths, covering the countryside with from twelve to twenty meters of debris.

Because many masses of mud flung directly from the crater fell to earth close by, some villages were spared while others near them were destroyed. The continued forward movement of this mass of rubble can be compared to the movement of a rock slide. The sharp edges on the rubble prove that it was carried along in the muddy water without being significantly abraded.

On January 5, 1823, the streams on the west and southwest sides of Merapi Peak rose to an exceptionally high level and, with a mighty roar, surged through the deep gorges of those same steaming rivers choked with ash, sand, and rock rubble. These mud rivers overflowed their banks many times.

During larger eruptions on Kelut Peak, mighty torrents of mud flowed forth. The water was yellow and seemed to contain large quantities of sulfuric acid; after it dried up, barren expanses of sand were left behind.

The navigable river Kali-Brantes, into which all the streams from Kelut Peak empty, rose mightily and carried along such a mass of uprooted or broken trees, dead buffalo, wild bulls, monkeys, turtles, crocodiles, and fish that a large bridge was swept away. The water was black, tepid, and stank of hydrogen sulfide.

This description is sufficient to give an idea of the severe devastation caused by present-day volcanic activity, running its course, for the most part, unseen by human eye.

Going on to fossil material, we should mention that the extinct giant turtle of Tenerife [Canary Islands], recently described by Ahl, was found in a Tertiary bed of volcanic tuff. The description given by Reiss (132) of the geological situation at a mammalian bone locality in Ecuador is also interesting:

The most important locality is the Chalang gorge, where the remains of fossil mammals are found piled up by the thousands. The richest layer of tuff may be the upper one, which is also characterized by pronounced salt efflorescences. It is difficult, if not impossible, to remove the bones directly from the tuff. The stone is hard, tenacious, and resists the blows of the hammer; the bones, however, are easily broken. It is therefore advisable to gather remains that have been washed out by rain or to make do with those on which subaerial agents have already started to work, and there is no lack of these. In the bed and on the banks of the stream, skulls are laid bare; along the steep walls of the ravine, the long bones stick out like coat hooks, and many are found loose between large boulders at the bottom of the ravine. Extraction of the bones has been made especially difficult because limestone, precipitating out of the Cangahua-like tuff, has often been deposited on and in them. Weathering proceeds fairly quickly, however, and in dry weather the surface of the Cangahua turns to dust and crumbles away. When it rains, water running down the walls continuously dissolves and washes away some of the tuff. Furthermore, the stream is also constantly eroding its banks, causing large masses of tuff to fall. Many of the bones are broken; still others, especially skulls, are well preserved. But a complete, articulated skeleton has never been found, and even with the skulls, the lower jaw is always missing. Wolf, who was the first to call attention to the Punin locality, was luckier. (Reiss quotes now from Wolf's report:) "In several places, especially in the ravine of bones at Punin, near Riobamba, there are thousands of horse and mastodon bones. From the lowest bed of tuff in this locality, I dug an almost complete skeleton, proof that the bones are found where they were first deposited. All the foot bones including the claws, as well as the upper and lower jaws and even the shell of the armadillo *Dasypus,* were found in a natural position and articulation." Nevertheless, such finds remain the exception. He continues: "The stream that ran through the Chalang gorge, near Punin, had cut through the volcanic tuff down to the underlying, nonvolcanic rock. Sandstone, quartzite, and silcrete make up the substratum upon which the great thickness (up to two hundred feet) of tuff rests. The lowest, and therefore oldest, layer of this tuff, recognizable at a distance by its efflorescence of saltpeter, is completely filled with mammal bones. The hills surrounding Punin once formed the shores of a lake that covered the whole Riobamba plain. Here, favorable conditions allowed the accumulation of an infinite number of bones. The remains of extinct animals such as *Mastodon andium [Cuvieronius andium]* are found with the bones of living species; therefore, the volcanic tuff must belong to the Quaternary. Because the bone-producing tuff is the oldest in the area, we can conclude that the activity of our volcanoes came relatively late and barely goes as far back as the Tertiary."

Quite a few cases of catastrophic fish kill must have occurred in connection with volcanic activity. In the last week of January 1925, millions of fish killed during a submarine volcanic eruption were washed ashore at

Wales Bay near Cape Down [South Africa]. The swath of dead shark, sole, cod, and other fish stretched out over two miles and was in some places two feet thick. The mass death of shark observed in 1902 on the coast of Somaliland was at first completely puzzling, but was evidently also connected with submarine eruptions. History tells us of the mass death of fish in the Gulf of Naples where, on July 16, 1794, a stream of lava from Torre del Greco poured into the sea (Abel 4). Johannes Walther (171) reports that on the day after the appearance of the volcanic island San Ferdinandea [Mediterranean Sea, ephemeral], vast numbers of dead fish were seen floating in the ocean, mostly pipefish, *Syngnathus anquineus.*

In the upper Pliocene on the coast of the Caspian Sea, numerous dolphin remains occur in volcanic tuff. According to Abel, the Caspian dolphin was completely exterminated at that time. Recurrent volcanic catastrophes in the interior of what is now the United States have made an exceptional contribution to the excellent fossil record pertaining to mammalian evolution during the Tertiary. Abel comments on the destruction of that fauna by volcanic eruption as follows:

Many of the North American tertiary formations are composed of volcanic tuff blown by the wind over forests and wide, shallow seas. Between these repeated periods of ashfall, lacustrine beds were formed. The formations of the North American Tertiary are named after isolated Tertiary basins, which at different times were home to great herds of mammals. The Bridger Formation includes, in all, five horizons of varying thickness: 60, 140, 90, 100, and 150 meters. In the second horizon, remains of thousands of animals have been excavated, indicating destruction of the entire biota by a mighty ashfall. The Santa Cruz Formation in Patagonia [Argentina] is a similar thick layer of volcanic ash.

Forty-two years ago, the fateful eruption of Krakatoa [Indonesia] took place. From May 20 to August 26, there was one eruption after another, and finally, two colossal explosions blew half of the almost one thousand–meter peak into the air. The islands of Krakatoa, or what was left of it, and Verlaten, lying opposite, were buried under two hundred feet of hot ash. Sebesy Island, in the vicinity, was also covered. On the first two islands not a single living thing survived. New colonization came from parts of Sumatra and Java, which frame the Sunda Straits, and from the smaller islands lying between them. Ocean currents brought seeds from Sebesy Island, Java, and Sumatra, and today forests, thickets, and meadows completely cover both islands. Most of the animals that live there now are flying creatures. By 1918, there were still no land mammals. By 1921, rats of the kinds found on Sebesy had spread out over the whole island. Two reptiles, a python and a monitor lizard, both very good swimmers often found far out at sea, reached Krakatoa in 1908. Two kinds

of geckos, probably washed ashore with driftwood, were observed in 1921. In addition, the following were found: eleven species of flightless insects, four millipedes, seventy-three species of spiders, land crabs, mollusks, and earthworms. All in all, a very interesting example of regeneration.

2. DEATH DUE TO GASES

Death from poisonous gases is often connected with volcanic activity. The best known instance is probably the *núee ardente* of Mt. Pelée [Martinique], which, with its temperature of one thousand degrees, killed twenty-eight thousand people within a few minutes.

Emissions of carbon dioxide are very dangerous and instantly fatal. Filhol (47) says that carbon dioxide from wells in southern France explains the extinction of life during the formation of the Paleogene phosphorite at Quercy. In the case of many a bird carcass, he was probably right. The Limagne [France], a once-volcanic region northeast of Puy de Dôme, still produces two hundred thousand kilograms of carbon dioxide daily; the gas well at Montpensier alone accounts for six thousand of that. Because of the gas, the pond at Montpensier is an animal trap still today. Methane and petroleum often accompany carbon dioxide in wells, and hydrogen sulfide usually does. However, not all emissions of hydrogen sulfide or carbon dioxide are of volcanic origin; for example, when siderite is changed to limonite by the action of oxygenated water, carbon dioxide is set free.

Accumulations of carbon dioxide are scarcely noticeable and are thus all the more dangerous. A few years ago, on the Finkelkule near Salzgitter [West Germany], a mining tunnel was constructed in the stratum underlying the Neocomian iron ore deposit, here composed of Posidonia Shale from the Lias. During construction, the sloping tunnel filled to a horizontal level with carbon dioxide; people were overcome by the gas and were resuscitated only with great effort. Carcasses of all kinds of songbirds and other animals accumulated quickly around the deadly site. Nearby flows a carbon dioxide well, which has also been exploited commercially.[4]

---

4. In the same way, we often find dead birds and mice in areas around the mofettes on the eastern shore of Lake Laach [West Germany]. Death from carbon dioxide is not poisoning but true suffocation due to oxygen deficiency of this heavy gas; its presence is so hard to discover that in the well-known grotto of Cane, near Pozzuoli [Italy], people use dogs to detect it. Curious are the unique explosive eruptions of natural gas on the Apsheron Peninsula [USSR] where, in 1902, a methane spring in the province of Kir Mateu, west of Zabrat, suddenly erupted and burst into flame, wiping out a herd of sheep. A. Tobler (*Topographische und geologische Beschreibung der Petroleumgebiete bei Moeara Enim, Südsumatra,* 1906) made a curious observation:

(*Continued*)

Fleeing animals springing onto muddy ground may be overcome and killed by the swamp gases released by their activities. The increase of carbon dioxide brought about by the growth and death of algae is devastating; such a bloom in fresh water is disastrous for fish. The massive die-off of fish in the Bay of Callao [Peru], in 1853, can probably be attributed to poisoning by hydrogen sulfide. The bottom of the sea there is covered with putrid, bluish black mud, and such die-offs had already happened repeatedly, though to a lesser extent. The often-observed mass death of fish in Walvis Bay in Southwest Africa is attributed by Stromer von Reichenbach (158) to poisoning by gases that sometimes escape from mud springs. It is difficult to determine just how deadly the gases at the sites of natural oil outlets are. The smell of the gases attracts coprophagous insects as well as vultures. In a stagnant pool containing petroleum near Lüneburg [West Germany], seventy-eight dead dung beetles were found. Where the petroleum flowed out they swam in the sticky mass, and when it dried up they crawled into mud cracks as much as seven feet deep (see Wiman 176).

Fresh grass submerged by flood waters seems to ferment, and the gases from this process cause the mass death of fish, as can often be observed in central Germany during flooding. Death from poisonous gases is also possible beneath a layer of ice. In Sweden, according to Wiman (176), they have a special word—*kraffisk* [winterkill]—for fish that have suffocated from lack of oxygen. Abel (7) has expressed the point of view that the remains of herds that have died during prairie fires need not bear traces of fire; the animals could have been overcome by fumes while fleeing.

3. DEATH DUE TO GRASS, PRAIRIE, AND FOREST FIRES

Devasting fires can wipe out animal life over a wide region and can even lead to complete displacement of the biotic community. The effect on wild herds of the prairie fires one used to see on the African veldt has been well described. These fires caused huge herds to stampede; the frantic animals turned to flee in whatever direction escape seemed possible. (Stampedes—the panicky flight of herds—can also have other causes: grass fire, thirst, lack of food, pursuit by wolves and other predators, criti-

---

The water at the locality had a prickly aftereffect caused by bubbling gases. The large forest animals—elephants, rhinoceroses, tapirs, and deer—drink such water with gusto and always return to this watering place. From miles away, well-trod elephant trails lead to the "spa," merging as they near it to form a dense network of paths one to two meters deep in the immediate vicinity of the spring.

We know of submarine gas eruptions in the Caspian Sea, on the coast of Burma, on Borneo, near Galeota Point on the southeast coast of Trinidad, in the Gulf of Paria between Trinidad and Venezuela, on the coast of Peru, and near Baku [USSR].

cal weather conditions, and above all, biting flies and mosquitoes. They are an important factor in the preservation of vertebrate remains because accidents often happen during their course. For example, the animals may encounter uncertain ground or run out onto ice; they may try to swim across large bodies of water or plunge into rivers, where they might be caught in eddies or perhaps prevented from getting out by banks that are too steep.) Wiman (176) thinks the effects of high water after a prairie fire are more important than any other factor for the accumulation of bones. In 1924, on the coast of Riviera Beach in Kleeberg County [Texas], near the Mexican border, I observed a great many turtles killed when the salt grass burned. The effects of grass fires are well known to vultures, and often the best way to attract them quickly is to set a very smoky fire.

During the panicky, frenzied flight caused by brushfire, even large animals may become stuck in swamps; thus, as so often happens, death comes from a combination of causes. Wiman (176) decribes the aftermath of a forest fire in Sweden, in 1913, during the course of which a large wooded area in the western part of the country went up in flames. The larger wild animals ran in panic and terror against the wind to a bog, where they were safe. The smaller animals, however, could not outrun the expanding fire, and carcasses of mice, hedgehogs, otters, and adders were found in the bed of a dried-up creek where they had sought shelter in vain. In Africa, the bush fire is also used in hunting, as Waibel (168) reports:

> Everywhere now the tall, dry grass is set afire and burned by the natives. Every evening the fire rolls in glowing red lines across the plains or snakes up the sides of mountains. Here, in this simple way, the land is cultivated by the native; the tall grass is reduced to fertilizing ash, and room for new grass and fields is obtained. The dry grass burns quickly, and the fire line moves speedily along. Lizards, snakes, and countless crickets are flushed by the flames and fall victim to the swarms of predatory birds that gather at every grass fire. These fires are seldom dangerous to humans: with enough foresight, people can always avoid them, or can jump quickly between the burning stalks to safety. For the frightened animals, it is different. Antelopes, especially, are often surrounded by a circle of fire and then slaughtered in huge numbers by the natives.

In October 1927, according to reports from Sydney, brush fires raged on the coast of Queensland, Australia, such as had never been seen before. Within a relatively short time, they covered an area 160 kilometers long and 15 wide. The city of Brisbane was cut off from contact with the rest of the world, and the fire's glow could be seen many miles out to sea. An area of 2,400 square kilometers does not burn up without killing untold numbers of animals.

In the effort to explain the origin of the enormous collection of *Stenomylus* bones in the area of the old streambed of the Miocene Nio-

brara River, Abel (4) suggested that prairie fires during long periods of drought had driven great herds of animals to the shores of the river, where they were felled by fumes.

4. DEATH BY DROWNING

We have seen that when carnivores pursue herbivores, the carnivores can easily chase their prey into bodies of muddy water from which escape is impossible. Wiman (176) reports that in the province of Jemtland, in northern Sweden, a herd of four hundred reindeer, harried by wolves, ran around crazed for days and nights and finally plunged into a lake, where they drowned. It happens not infrequently that birds of prey swooping down on their victims hit the water instead and cannot save themselves. Flöricke often found hawks that had died in this way.

Such animal catastrophes are also often connected with the great migrations, our knowledge of which must be gotten from older literature, since human culture has gradually reduced the number of such occurrences considerably, and scientific interest in "catastrophe" has diminished somewhat too much. Nevertheless, as late as 1902, during the great plague of mice in Rheinhessen [West Germany], it was reported that a mighty procession of these creatures swam across the middle Rhine. In 1727, in Astrakhan [USSR], an enormous number of brown rats drowned while swimming across the Volga during their migration out of Asia into Europe. In 1847, vast numbers of squirrels trying to swim across the Jenissei [USSR] were carried off by the current, and all drowned together. (According to Wiman [176], squirrels migrate singly and their movements go almost unnoticed. Gatherings in such numbers happen only when an obstacle is encountered.) On the Gulf Coast, I repeatedly observed nocturnal animals—opossum, raccoons, and ring-tailed cats—disturbed and apparently hunted, jumping into the river from overhanging branches and dying, even though they knew how to swim. Concerning the migration of lemmings (somewhat in doubt today), it has also been reported that the animals have an irresistable urge to plunge into the water, and so drown by the hundreds of thousands. Concentrations of bleaching carcasses of these animals, found in recent times in a large bay of the Lulaeaven River [Sweden], have been described. Older reports maintain that fishing boats in the Gulf of Bothnia often became so filled with them that they sank.

Recent skepticism regarding lemming migration definitely seems unwarranted. In 1910, at Tromsö [Norway], B. Högbom (Wiman 176) saw a pack of swimming lemmings two to three kilometers long. A. Högbom (Wiman 176) observed a dense column of a kind of vole. Wiman (176) reports how he traveled for an hour through a pack of swimming lemmings, and he believes that "the countless lemmings dead in the open sea have great paleontological significance." Many also drown swimming across rivers and freshwater lakes. On the shores of a lake in Jemtland,

Wiman found an embankment made entirely of dead lemmings, one more example of concentration of remains by washing ashore. Migration, mass death by drowning, and the embedding of land mammals in marine strata are here closely connected.

According to Nehring (118), the above-mentioned lemming, *Myodes obensis* Brants [*Lemmus sibiricus* Kerr], migrates regularly across the Samoyed tundra [USSR]; from the end of May until the middle of June, these animals cross the Ural Mountains to the plains and spread out over the tundra as far as the White Sea and the treeline. Wiman (176) does not accept lack of food in the mountains as the cause of the migration. He thinks, rather, that an epidemic sickness causes "a kind of craziness." It is obvious that epidemic infections often take place during the massing together of animals for migration. However, the reason for the migration must, in my opinion, be biological and not pathological. Even in lemming years, we find only a few dead individuals in the forest, and these eventually disappear completely.

We shall describe later how disastrous it is for birds when oil befouls their plumage. Destruction of the feather insulation of a bird's warm body causes it to freeze quickly. Wiman (176) observed that scavenger gulls soaked during a skirmish are doomed. They swim to shore, stumble around all day, and finally die from hunger and cold; without their natural thermal protection, they do not return to the cold water. Because this description is of a fight over a whale carcass, the possibility that the feathers became defiled with oil is not excluded.

In the carcass assemblage at Smithers Lake, bird remains are also found strewn here and there among the debris washed ashore, but never in heaps. They are mostly the remains of wild ducks whose plumage was damaged by drifting masses of wood and carcasses, causing them to freeze to death. The plumage is so badly mauled and stuck together that it is hard to recognize the remains as those of ducks. Some examples, whose position and appearance are typical, are shown in plate 2, figures B and C.

There have been a great many reports of mass death among migratory birds, especially in the area around the Black Sea. The cause of death usually involves complete exhaustion. Wiman (176) gives an interesting example provoked by a change in wind direction: "The wren flies only with the wind. A large flock of wrens had gathered in southern Öland [Sweden] and was awaiting a favorable wind. Finally it began to blow from the north, and the whole flock flew away. Then the wind changed, and somewhat later great numbers of drowned wrens were washed ashore."

There is a curious connection between the sucking, irresistible force of waterfalls and mass death by drowning. For example, swans alighting on a river after a long migration are prone to encounter this type of disaster (see Wiman 179; Cole 33; and Fleming 48). The roar of the approaching falls seems to unsettle the creatures, and the mist further confuses them;

consequently, at the predictable spot below the waterfall, dead birds accumulate. They are almost always water birds, or at least good swimmers such as swans, geese, ducks, and coots. According to old French reports, at Niagara Falls, during the months of September and October, such quantities of birds were found that the garrison of the nearby fort lived almost exclusively on them for a long time. In addition to birds, carcasses of fish, deer, bear, and other animals were found; all had been overcome while swimming by the force of the water rushing toward the falls.[5]

5. BOGGING DOWN IN MUD

The phenomenon whereby calcium-rich, clayey prairie soils dry out until they are hard as stone and then become completely saturated again is widespread. Pockets occur, probably connected with this constant dry-

5. Salomon (in *Geol. Rundschau,* 1926) has recently and in a noteworthy way compared the distribution of the bones of drowned vertebrates in the sand of Mauer with the Rehbockian whirlpool effect:

> Because I have been occupied since 1901 with the retrieval of fossil bones from Mauer on the Elsenz [West Germany], the famous locality of *Homo heidelbergensis* [*Homo erectus* d'Europe], it has long struck me that the skeletal elements of the large mammals found there almost never showed the effects of transport, yet were not found articulated. Very rarely do we have a skull with its lower jaw. Only a single complete skeleton—from a small cervid—has appeared. Furthermore, pectoral girdles, pelvises, and legs are almost always found alone. I have reported these facts and explained them by saying that the locality lay in an old loop of the Neckar, and that the carcasses swirled around so long in the eddies that the individual parts rotted off. An especially characteristic instance of this thesis and supporting evidence for it is the following:
>
> As far back as the 1860s, the Zoological Institute of the University of Heidelberg had part of a deer skull with one of the antlers attached. Forty to fifty years later, when the sand quarry operation had long since moved to another location, the other half with the other antler was found. Both parts had been so little abraded that they could be fitted together.
>
> According to Rehbock's interesting observations, the locality at Mauer was once not an ordinary whirlpool but a bank eddy of considerable length.

We have Pfizenmayer to thank for a description of how a creek carried parts of a rhinoceros skeleton, each part according to its weight, over a distance of about a kilometer:

> As I was walking along the bed of a dried-up creek near Verchojansk [USSR], I found first the femurs of a rhinoceros, then several larger and smaller bones, some with tendons and ligaments still attached. As I went upstream, I kept finding more pieces of the skeleton: leg bones, scapula, ribs, vertebra, an atlas on a tree root, and finally, in a puddle, part of the pelvis and the upper part of the skull of a *Rhinoceros tichorhinus.* No more bones were found upstream beyond these, the heaviest parts of the skeleton. The bones had been spread out for nearly one kilometer.

ing out and soaking, where sticky, plastic clay and lean, often quite thin, sandier prairie soil interact. At the contact zone there is often a black band where the sand is thoroughly mixed with humus. The clay has here attracted adsorptive humus substances to itself. The upper parts of the clay are usually yellowish or grayish yellow, whereas the parent rock was often red. There is a zone of reaction here where limonite concretions form chemically. Carbonate concretions owe their formation to similar processes; they are usually found at the edge of the prairie next to wooded floodplains.

Various surface forms resulting from this soil structure cover wide areas of Arkansas, Louisiana, and Texas. In some places, the sticky subsoil has a tendency to swell and pierce the thinner surface layer, or the sand in the upland flats of the coastal plains forms characteristic round dunes, called "mounds." Another consequence of the uneven structure of the two uppermost meters of the earth is the characteristic shallow prairie lake; its northern shore is always flat, but the southern shore is steep and reeflike, reflecting the strong effects of the north wind, which causes it gradually to migrate. These temporary bodies of water are exceptionally important for the propagation of turtles and a number of interesting crayfish, which, when the water recedes, bury themselves deep in tunnels and build peculiar mud towers at the mouths of their burrows. Where irrigation is practiced, these creatures do quite a bit of damage to the dams in the rice fields. This microrelief is especially important for the distribution of seed, the arrangement of vegetation and ant hills, and many other biological events and relationships. Flooding brought on by sudden rains causes the formation of strandlines of seeds and dry plants; these high watermarks extend more or less widely over the sand hills and can be an indicator of past maximum and minimum precipitation levels. We can best understand how suddenly the prairie can become soaked if we realize how much precipitation can result from one rainstorm. The prairie can also dry out so thoroughly that deep cracks yawn and travel on horseback becomes dangerous. The repeated breaking open of these cracks during drought and swelling shut in wet times also causes interesting minor forms.

On the prairie, when heavy precipitation follows a drought, some places swell more easily and quickly than others and may lead animals to their destruction. There are also places soaked year round where one can see individual animals sink in and slowly die without causing much concern to man or beast.[6] One person alone cannot pull a cow out of the mud; to get

6. Livestock sinking into mud is not exactly a rarity here, either, but its monetary value evokes a more humanitarian response. As an example, here is an item from a 1926 newspaper:

Braunsroda [East Germany]. (The cow that sank in the mud.) A local farmer was taking his cow to the neighboring village to be bred because at the time there was no bull in his own village for stud. On the way back, want-

(*continued*)

help would cost more than the animal is worth and is usually not done. Over and over I have seen this kind of insensitivity in people who live in hot countries, and recently, Abel (4, p. 137) vehemently expressed his feelings about a similar incident.

There is an additional mechanical factor: quicksand forms very quickly because water penetrates the pore spaces between the sand grains and eliminates friction. The resulting consistency is not easy to alter. However, if there is a mechanical disturbance causing water to penetrate the spaces between the very finest particles, then colloids are formed against which the greatest strength is powerless. Compared to lean, colloid-poor clay, fat clays quickly become greasy in wet weather because they are rich in colloidal compounds. We need only recall how streets still tolerably passable for one person were churned to bottomless mud when a column of soldiers marched by; places where a single animal can still pass become suddenly fatal when the whole herd goes through. Then, as the earth dries out, it holds fast to whatever is mired in it. Remember that a modern brick factory operates on the principle of stirring some of the components to a fine, colloidal mud soup, and then mixing in the rest of the ingredients, in pieces. It is characteristic of the Gulf Coast that when a norther hits, automobiles on all unpaved streets and dirt roads suddenly come to a standstill and are hopelessly stuck, and racing the motor to free them is, after a certain point, the surest way to remain stuck. The spinning wheels mix earth and water into a bottomless mass of muck. It is perhaps really not as out of place as it may seem to compare stalled cars on the prairie with heavy dinosaurs that suddenly encountered ground that could no longer bear their weight and became mired. Death due to the legs becoming trapped in sticky mud is most common throughout the prairie.

Two other factors favor this type of death: First, a number of vicious, biting flies choose the line between hoof and skin to attack. Under certain weather conditions, the animals, tormented to the point of madness, can become so beside themselves that the whole herd stampedes, not stopping until it comes to a swamp or a shallow body of water, or even just damp earth; in such moist environments, they can rid themselves of their tormentors and cool the inflammation of the bites. The other danger lies in cold and exhaustion. When a norther roars across the prairie, herds turn away from the wind and head south; if they encounter no obstacles, they will continue until completely exhausted. Consequently, after a deadly storm carcasses are found piled up along the wire fences at the southern boundaries of property and on the north sides of train tracks running east and west.

---

ing to take a shortcut, he led the cow across a stubble field still muddy from winter moisture. He had scarcely gone fifty steps when, to his horror, the cow sank in up to its belly. He had to return to the neighboring village, find a dung cart with two oxen and several people to help; they loaded the cow onto the cart and took it back to its warm stall.

Let us look at a dead cow mired in a place that normally would hardly be considered an animal trap (pl. 2, fig. D). The extremities are fixed in the mud in a characteristic way. The foreleg points forward, the hind leg to the rear. The position of the head is also characteristic. The pelvis shows the extremely typical evidence of feeding by vultures.[7] These exhausted animals often live three or four days before finally dying, usually from hunger. The same phenomenon appears in a mired cow that died in December 1924 and was photographed on October 10 of the following year (pl. 3, fig. A).

During the devastating norther that killed roughly 1.25 million cattle in a few days, the herds from the wide salt-grass marshes shown in the background of the picture turned toward the coast. Where the clumps of salt grass meet the bare ground at the edge of lakes and at the Intracoastal Canal at the east end of Matagorda Bay, many animals remained as shown in plate 3, figure A, with their legs completely immobilized, mired up to their bellies. Under the burning sun the hide dried out hard and parchmentlike. Shreds of it stuck to the head and belly, and the stomach contents welled out. A few animals succeeded in crossing this sticky, muddy place and, still fleeing the icy wind, pressed on into the water of the canal. Hundreds and hundreds of their carcasses lay on the bank; because those in the water completely blocked the ship channel, they had to be pulled out.

Such death due to miring is also entirely possible in the middle of arid basins. Salt and alkali flats can have a completely hard, dried-out crust and still be hygroscopically damp and yielding underneath. Heavy animals break through these places as through a crust of ice and cannot free themselves. Stappenbeck (155) was able to observe this repeatedly. The sudden increase in adhesive strength and viscosity of the sediments creates an action of force that causes the formation of inescapable traps. The giant marsupial described as *Diprotodon australis* Owen from Lake Callabonna [Australia] obviously met a similar death. Stirling and Zietz (157), Abel (4), and Woodward (182) have researched the burials of these rhinoceros-sized marsupials.The skeletons were partially exposed and coated with travertine. Without exception, the feet of complete skeletons were stuck deep in what was once mud, while pelvises, backbones, and skulls lay higher up. The animals certainly died on the spot in the manner

7. In addition to vultures, whose help is enlisted by Parsis and Tibetans in the disposal of their dead, there are many other scavengers. In China, the dead left lying out are eaten mostly by pigs, and our wild pigs also eat carrion. Hyenas, with their strong jaws and sturdily built forequarters, are also scavengers, as are coyotes, wolves, jackals, foxes, dogs, and Norway rats. Schomburgk has even photographed leopards feeding on carcasses. Many rodents, especially mice and squirrels, gnaw bones. The evidence of gnawing teeth is particularly noticeable in bones and shed antlers lying in the forest and is also found in fossils. Gulls, crows, storks, ants, crocodiles, and most fish, especially carp, lampreys, morays, and eels, are also scavengers.

described above. Salt pans and saliferous clay soils are found throughout arid areas. The North African *sebkhas,* the *takyrs* in Turkestan [USSR], and the *kewirs* in Persia [Iran] are examples; in these places a salt bank lies beneath a ten-centimeter layer of dry clay. Underneath is a moist clay, of softer and muddier consistency. Similar phenomena apparently occurred in the dinosaur localities in the Tendaguru beds of German East Africa [Tanzania]. In some places, bones of extremities, pelvises, and shoulder girdles were found, but almost no backbones. Sinking into mud seems to have been the main factor here. The position of the legs of many North American dinosaurs attests to the same situation. Large ungulates make deeply incised trails from which the vegetation disappears completely, leaving the earth bare. If the trail crosses a depression that fills easily with water, animals can become trapped at that point. The American bison made many of these trails, and cattle do it still today.

Austrian lignite deposits present partial skeletons, which by their position are suggestive of miring in mud. Mammoths, too, often fell victim to treacherous ground.[8] Similar discoveries have been made in the Patagonian Miocene and also in the younger formations on the pampas, where the position of many giant sloth skeletons attests to their destruction by sinking into mud when drought compelled them to seek water in shrinking rivers and lakes. There is also geological evidence of animals having become mired in bogs, not only in Tertiary lignite, but also in Ice Age peat bogs. There, it was usually not a question of mass death, but of dangerous places where separate accidents happened over and over. Wiman (176) mentions springs in Lapland where horses often sink into the soggy ground and die. He reported having come across a dead moose in a bog near Upsala [Sweden]. Siberian mammoths sank into mud streams in tundra bogs. Carcasses of giant elk in the Irish peat bogs and *Bos primigenius* in the Vig peat bog in Denmark should also be mentioned here.

6. STOMACH CONTENTS

The sinking of heavy animals into sodden ground is often directly related to food consumption. During prolonged drought, animals search for food at the edges of bodies of water and in other areas that under other weather conditions would never be accessible to them. When everything dries out, however, the animal is attracted to places that have retained at least a little soil moisture, where browsed vegetation can still renew itself to a certain extent. It is exactly these places that become traps—patches of

8. For this reason, the Yakuts are absolutely convinced that Trajonscheitan, the prince of hell, dwells in vast subterranean halls deep beneath the quaking, trecherous, mossy surface of the "sinking tundra," an area animals avoid (see E.W. Pfizenmayer, *Mammutleichen und Urwaldmenschen in Nordost-Sibirien,* Leipzig, 1926).

bare ground where, at the onset of a downpour, the subsoil quickly becomes saturated and treacherous.

When grass becomes dry and sparse during a dry spell, an almost inconceivablely large, indigestible mass of cellulose accumulates in the stomachs and intestines of ruminants. Plate 3, figure B shows the carcass of a cow mired in damp ground; its right flank has been torn open from the rear by feeding wolves. The left flank has burst, and the contents of the stomach fills the whole body cavity and spills out to cover the ground for the entire length of the carcass. Note the composition of the mass—caked remnants of dry, tough prairie grass—and its large size compared to the size of the dead animal. It requires no great explanation that following mass death, after which the carcasses lie around for a whole year, the number of wolves, coyotes, and other scavengers increases to an unusual degree, as was clearly seen in Texas in 1925. I noticed it myself near the Big Creek oilfield, where, at one such accumulation, two old wolves, *Canis nubilus* [probably *Canis latrans*], were killed and eight juveniles captured.

It seems as though carcasses of animals that died in mud or quicksand are especially apt to leave the remains of their stomach contents behind. Remains of fodder containing members of the genus *Salsola* and the families Amaranthaceae and Nyctiginaceae were found with the skeleton of *Diprotodon* from Lake Callabonna, in Australia. Animals that die in quicksand formed by unusually high water levels may become quickly mummified when the water recedes and the sand becomes dry and loose again. It is entirely possible that the *Trachodon* [*Anatosaurus*] carcasses of North America are the result of this kind of death. In these animals, too, remains of stomach contents can be seen. During preparation of the specimens housed at Frankfurt, as described by Kräusel (95), compact, gray, earthy masses composed of a little sand and numerous plant fragments, including needles from conifers and twigs from deciduous trees, as well as many small seeds and fruits, were found inside the carcasses. These masses cannot be merely washed-up, rotting plant material because they contain no pollen, saprophytic fungus, or the like. The composition of the masses indicates that the animals grazed widely and did not limit themselves to low ground vegetation.

The stomach contents of the *Hybodus hauffianus* in the magnificent exhibit in the Stuttgart [West Germany] natural history collection [Staatliches Museum für Naturkunde], described by Schmidt (143), caused quite a sensation. The stomach area was filled with a ball composed of the remains of approximately 250 belemnites, so closely packed that their arrangement corresponded exactly to the most economical use of the space.

Piles of disgorged belemnites have also been found, but Schmidt (143) goes too far when he attributes these "belemnite battlefields" directly to feeding sharks. These submarine areas of accumulation and their concentrations of fossils are caused by more turbulent water and occur in places where the turbulence has smoothed the seafloor. The result of such smoothing out of the profile is that the water moves even faster.

Compsognathus (Abel 3; Wagner 167) has the remains of a small reptile lying in its stomach. In the Mansfeld syncline, Kupferschiefer fish containing *Productus horridus* have been found. Juvenile ichthyosaurs that had been eaten were found in the Posidonia Shale. The Banz saurians from the same shale are always accompanied by gastroliths. *Diplocynodon* droppings from the Geiseltal lignite also yield gastroliths. The stomach contents of mammoth and rhinoceros carcasses have often been described, and fish scales have often been found near the remains of fish eaters.

7. DEATH IN QUICKSAND

This kind of death is similar to the one caused by the sinking of animals into gummy mud. Quicksand is a mixture of sand and water in which friction between the grains of sand is eliminated when water fills the spaces between them. Consequently, the sand gives way fairly easily under the pressure of a heavy object and then flows quickly together again. One can best learn about the origin of quicksand by reading Berendt (21) and Andrée (9). The most dangerous form appears when water is under hydrostatic pressure, although water may also penetrate loose sand horizontally; finally, quicksand is formed when masses of sand slide slowly into standing water. Notice how difficult it is to cross the mouths of small beach streams whose water is dammed up and must percolate through the sand.

What interests us most in this study is the formation of quicksand when water levels change. Even tides can produce this effect,[9] especially when water that has flowed across a flat beach ridge retreats. According to Andrée, the depth of the quicksand depends on the height of the beach ridge. The formation of quicksand becomes more extensive during abnormally high water levels. During storms, and especially during flood tides, the water level on our German coasts can rise three meters and more above normal; it is the same on all other coasts during heavy storms. At these times, water forced out of the ground hydrostatically can turn loose, dry, sandy places into dangerous animal traps, especially when the water level rises slowly.

In fact, the severe northers on the Gulf Coast always result in great loss of livestock in coastal quicksand. Cattle grazing on the salt marshes turn

9. Quicksand often forms on sand flats at ebb tide. Accidents of this kind are reported all the time; here is an example from a recent newspaper report: "Tragic death of a fisherman. It was reported from Paris that a fisherman and his friend were going to the Bay of Mont St. Michel. Taking a shortcut, they crossed a large sand flat where the fisherman, unnoticed by his friend, suddenly sank in without a trace. The search for the body has so far been fruitless."

south away from the cold wind and head for the much warmer water. The exhausted animals become careless, start to doze, and get caught in the quicksand. Whole herds may meet their deaths in this way. After the storm, the abnormally high water recedes, and it is strange indeed to walk across this field of carcasses where for the moment the only things to be seen are oyster shells strewn about and the protruding horns of cattle casting sharp shadows on the pale yellow sand.

This course of events was observed on the open Gulf Coast during the same snowstorm that caused the catastrophe at Smithers Lake. The freezing, starving cattle fled into the water to escape the northerly wind. The carcasses often sank in so completely that not much more protruded than the nasal bones and the tips of the horns (pl. 3, fig. C). The following year, the carcass assemblage was gradually exposed, due less to waves than to wind, which dried the sand and carried it off (pl. 4, fig. A). Sand-dwelling beach crabs had already removed the flesh from most of the skeletons, and there was often a whitish calcicrust deposited on the bones, especially on the pelvises.

These cattle carcasses, partially exposed by the wind, were in a curious position—they lay on one side and were often fairly sharply curved, so that the head lay near the pelvis. Often, the pelvis no longer lay in a normal relationship to the backbone (pl. 4, fig. B). The skull usually lay on its base, but was sometimes also found in an oblique, lateral position (pl. 4, fig. C) or tipped over (pl. 5, fig. A). The pelvis was often in a dorsal position, seldom articulated. The spine was usually somewhat contorted, so that when the head lay on its base, the spine might be lying on its side, often bringing the pelvis up near the broken bones of the forelegs and scapulae.

If the sand was well penetrated by dry air, mummification also took place. The legs of such carcasses were often quite bent (pl. 5, fig. E; pl. 6, figs. A and B), as described by Abel (3) at Pikermi [Greece]. Plate 6, figure B shows a severely buckled carcass just beginning to be exposed by the wind. The hind legs lie almost parallel to the skull, which is turned toward the rear. Of the pelvis, only one socket [acetabulum] is visible. It lies with the ventral side vertically down. The vertebra of the neck are squeezed laterally. It is often interesting to see how the obstacles formed by the protruding bones produced lines of force in the turbulent sand. The activity of sand crabs has already been mentioned. Plate 6, figure C shows that they can also move parts of skeletons. The tracks of these old robbers can be seen next to the calf skull and the displaced maxilla. Hermit crabs are also efficient skeletonizers.

Vertebrate carcasses already lying on a beach are often washed up farther onto sand surfaces that turn to quicksand when the water level next rises. Because of its weight, a carcass presses down on the sand, squeezing it out around the edges, and its bulk forms an obstacle to the receding water, compressing its flow. Soon, a ring-shaped erosional depression forms behind the carcass, and one or more drainage channels form in front of it, with the result that the carcass can sink in and be embedded relatively quickly.

Matthew (Abel 7) thought he could attribute the huge, individual-rich but species-poor vertebrate deposit of the Lower Harrison beds near Agate, Nebraska, to mass death in quicksand, a point of view contradicted, however, by Abel (7). According to Matthew, the Niobrara River, where the animals went to drink, flowed twenty-five meters higher at one time. The watering place was near a strong eddy, where the subsoil formed quicksand during dry spells. Ligaments and tendons of the animals that sank into the quicksand probably dissolved during decomposition, and the elements of the skeleton worked themselves down, precluding further transport, damage, or predation. Matthew explains the fact that only large ungulates such as *Diceratherium, Moropus,* and *Dinohyus* were buried by saying that a smaller animal would not have weighed enough to sink into the quicksand.

8. Becoming Mired in Mud Flats

Animals that have died after becoming mired in mud flats are similar in appearance to the quicksand carcasses previously decribed. In this instance, too, the dying animal often turns its head toward its pelvis so that the nose points backward. Consequently, the whole collapsed skeleton takes on an odd circular shape. The ends of the ribs become detached from the spine and point up, but the ribs tend to lie aslant or even almost horizontal. At low tide on the mud flat near Cameron, Louisiana, I came across such a circular, disintegrated horse skeleton; it had already been colonized by young oysters. One noticed, among other things, that the oyster shells were molded around the chewing surfaces of the teeth. During rapid siltation, as soon as these fast-growing mud oysters can no longer function, they die and the shells open; many dead animals thus become the foundation of a small oyster bed. Once my boat ran aground on the oyster-covered bottom of the large Calcasieu Lake, and during the night I heard the anguished cries of cattle from the salt marshes between the outlets. The next morning I saw several of them mired in the mud at the edge of the grass, slowly sinking in. On foggy days, accidents like this happen often. Humans, too, can suddenly sink into a soft mudbank right down to a shell bed, which usually provides a foothold.

9. Death in Crude Oil and Asphalt

Another mode of death, death in natural crude oil and asphalt, is similar not only to dying in mud and quicksand, but also to becoming icebound and drowning. The discoveries of bones in the California asphalt pools are well known; indeed, the famous tar pit at Rancho la

Brea is within easy reach of Los Angeles. It is less well known that in Central America and northern South America, prospectors sometimes find promising oil seeps by first spotting a skull or other bones from a saber-toothed tiger protruding from the forest floor.[10] In American publications (Scott 149), we often come across artists' conceptions of a typical scene in a California asphalt swamp: An elephant has become mired and has dropped to its side. A saber-toothed tiger stands on it and drives two giant wolves away from the carcass into the asphalt swamp, where they, too, fall victim to the treacherous ground.

From the list of fauna at Rancho la Brea, it emerges clearly that predatory birds predominated by far, especially if we also take the number of individuals into consideration. Following the odor of the pools and the carcasses in them—carcasses of red deer; horses; camels; *Paramylodon;* saber-toothed tigers, lions, leopards, lynx, and other wild cats; bison; elephants; prairie and dire wolves in particularly large numbers; foxes; moles and shrews; bats; squirrels; rabbits; skunks; pumas; badgers; two kinds of bear; Virginia deer; antelope; peccaries; tapirs; and mastodons— the vultures always located the most likely spots, and many of them became victims too. One particular assemblage contains thirty-three golden eagles and only three nonpredatory birds. Among the mammals, the ratio of predators to nonpredators is exactly even. The locality at La Brea shows that the golden eagle has long had the habit of swooping down on weakened or dead animals.

Because of the comprehensive nature of the entrapped fauna, animal traps in the form of oil seepages can provide important information on geographical distribution. For example, a peacock (Miller 112), known until then only from Asia, was found in the Pleistocene fauna of La Brea. It was, of course, not an Asian species, but a North American one, something between the South Asian peacock and the meso-American turkey.

If we assume that one of the attractions for mammalian herbivores was probably water, it is striking that wading birds have so seldom been found in these beds. The bird fauna of La Brea was anything but well balanced. The abundant fauna in the area of the asphalt lakes at the time of their formation made the lakes especially attractive to vultures, which often fell in and were preserved. There is doubtless a specific ecological relationship between the large scavenger birds and the large predatory mammals similar to that between the large herbivores and the plentiful vegetation they required for food. Herbivores killed and partly eaten by carnivores— in Rancho la Brea over three thousand *Smilodon californicus* (saber-toothed tiger) and even more dire wolves were found—are the natural food of large vultures. It is also obvious that the asphalt lake was more attractive to vultures than to other animals. Crude oil gases, often fatal even

---

10. Bones of large animals have also been observed in the asphalt deposits of Peru (see A. Beeby, *Petroleum Mining,* London, 1910, p. 104).

to coprophagic Coleoptera, seem to attract vultures, a bird indifferent even to the odor of the skunk. Nine different kinds of vulture, of which two are Old World types, occurred, and 28 percent of the rest of the brids were golden eagles (Miller 111).

Once carcasses are enveloped in asphalt, they cannot dry out and mummify. Shut off from the outside, the intestinal and putrefactive bacteria work from inside and in a short time liquify the whole carcass. In asphalt lakes today, we find animals whose sacrum, back, and cranium are heavily weathered, whereas the legs, sternum, base of the skull, and front side of the head are totally saturated with oil and consequently well preserved. Among the carnivores found at Rancho la Brea, juveniles and very old animals with worn, broken teeth predominated significantly.

The deposits at Fossil Lake, in Oregon, present a picture similar to that at the asphalt lake at Rancho la Brea. There, similar mammal remains, a few fish, and many birds were found. Whereas at La Brea, the solid tarsometatarsals constitute by far the greater part of the remains, at Fossil Lake, except for a few coracoids and scapulas, only fine bones were found, which are normally not well preserved, and tarsometatarsals occur extremely seldom. Miller (113) found a plausible explanation for this disparity: thanks to the buoyancy of the layer of air enclosed within the feather mantle (I myself think intestinal gases and the pneumatic bones are just as important), the bodies of dead waterfowl can stay afloat until prevailing winds blow them ashore. The bare lower legs are submerged the whole time and, consequently, more subject to decay and attack by water insects. These influences speed up dismemberment due to loss of the connecting tendons. The metatarsals, therefore, were strewn over a wide area, whereas the rest of the remains were concentrated on the shore of the lake. In skeletons of seabirds on the beach, the ligaments of the foot and lower leg dissolve first, often causing these bones to be missing. Only thus can we understand that the pneumatic coracoids were preserved, but not the solid tarsometatarsals.

The makeup of the avifauna of the McKittrick asphalt, also described by Miller (114), is completely different from and much more balanced than than that of Rancho la Brea. Some one thousand bird carcasses were divided among thirty-four species, twenty-eight of which still live in the area; three of the species live today only in other areas, and three are extinct. Whereas at La Brea the number of raptors was unnaturally greater than the number of nonraptors, at McKittrick the number of waterbirds—ducks, storks, and shorebirds—was twice as great as that of the rest of the species, and migratory birds were better represented than at Rancho la Brea. Gulls, loons and grebes, and pelicans were totally absent, as were herons and small species of vulture. [According to Miller (114), herons were present at McKittrick—TRANS.] The large forms found at Rancho la Brea—*Catharista, Cathartes,* Cathartidae, and condors—were not present at McKittrick, but there were other species of birds, attracted by carcasses of camels, elephants, and lions. The most common species was the golden eagle. These swamps were too far from the coast for oceanic fish-eaters to have been present. The difference between the two faunas is due less to the time difference than to the difference of geographic milieu. The McKittrick fauna was that of a marshland with shallow ponds but no large lake, and it was younger than that at Rancho la Brea or Fossil Lake. McKittrick, too, was a Pleistocene crude oil seep, but probably wider than the others. Wetlands and vegetation mask the deep animal trap, which claimed animals as large as a mastodon and as small as a mud-gathering swallow. Damage by weathering is seldom encountered. More broken bones occur there than at Rancho la Brea, probably because the crude oil that formed the asphalt was of a slightly different composition. The embedding medium is light in color and mixed with fine-grained mud, while the dark, gummy asphalt from Rancho la Brea contains grains of sand.

Oil slicks on water hold an unusual and fatal attraction for seabirds. The shining surface reminds them of a seething mass of silvery fish, and they dive into it. Their feathers become contaminated with oil, and when they try to get it off with their bills, it sticks even more. Finally, the feathers are stuck completely together, and the feather mantle fails. The bird can no longer rise, the insulating layer of air escapes, and the cold water comes in contact with the body; then the bird either freezes or starves to death. This kind of death can occur naturally, because there are also natural submarine oil seeps, as, for example, at the mouth of the Sabine River, west of the Mississippi Delta. The masses of asphalt that often wash ashore along the coast of the Gulf of Mexico, gluing together shells of snails and bivalves thrown up by the waves, attest to this fact. However, since most steamships have now been converted to oil, the killing off of seabirds has reached inconceivably catastrophic proportions. Only upon encountering a mass of the carcasses of these oil victims on a beach does one begin to grasp the enormity of the problem. The ornithologist Weigold, in Hannover, has raised his voice in favor of international measures to protect birds from this mass destruction. May his plea be heard!

10. DEATH DUE TO FLOODING

Seasonal flooding plays an interesting role in the life cycles of freshwater fishes. Antipa (17) has made important observations in the delta of the Danube River [Romania] in the practical interests of the fishing industry. He divided the fishes into three groups:

1. True baltafish [*balta* is the Romanian word for the small lakes found in the floodplain of the Danube—TRANS.], namely, bream, pike, tench, Prussian carp, European roach, perch, sheathfish, zander, reed carp, sicklefish, and so forth
2. Fishes that travel from one lake to another, namely, carp of all sizes

3. Fishes that migrate from the ocean to the Danube or the lakes, such
as (*a*) species of sturgeon (beluga, sturgeon, sterlet, sherg, fat, and
smooth), which enter the Danube year round; (*b*) mullet (Mugi-
lidae), which come in the summer from the Black Sea to lakes
Razim and Sinoe, and the lagoon (Zaton) near St. Georg; and (*c*)
turbot *(Rhombus maeoticus)*, fished in this area off the coast near
Sulina and in the area of Portita as far as Caraharman; the catch of
such fish depends more on ocean currents, wind direction, temperature, and so forth than on the water level of the Danube

The annual catch must be compared with the maximum high of the river
in spring and the length of time the flooding lasts. The carp catch in years
with low water levels—1898 and 1899—went as high as five million
kilograms; in the years 1905–6 and 1907–7, when the water levels were
high, the catch was much smaller; nevertheless, the size of the catch of
baltafish is always directly related to the water level of the Danube. The
decrease is not noticeable until the following year because actual reproduction takes place in the flooded area. Between 1895 and 1899, production rose and fell according to the water level of the Danube. In 1899, the
water was very low and yielded a catch of almost one million kilograms.
Because in 1899 the river carried so little water, by autumn the fish were
crowded together in small basins, where they were caught in larger numbers than was good for the survival of the population. Consequently, the
catch in 1900 was off, although the water level was high—3.53 meters. Production grew normally until 1902. In that year, spring floodwaters rose
only as high as the banks and remained only a short time. In the autumn,
the water was very low in the lakes as well as in the river. Then, when a
sudden frost hit, accompanied by snow, vast numbers of fish suffocated.
That winter, fishermen pulled out such enormous quantities of carp that
even old tar barrels and boilers were used to salt the fish in. Consequently,
the catch was seemingly high—5.5 million kilograms. In the next year,
however, the damage made itself felt. In 1903, the catch was only 3,813
million kilograms. Then normal conditions prevailed, and in the years
1907 and 1908, 7,739 million kilograms of baltafish were caught.

In the delta lakes, too, actual fish production grew in direct relationship
to the amount of water carried by the Danube. The shallow, food-rich
waters of the flood zone favor unusually fast growth of the offspring and
are, in this respect, an excellent environment. Adult fish spare no effort to
reach these suitable spawning grounds to carry out their reproduction. But
these waters may also pose great danger if either the water drains off too
quickly or the outlet is dammed. In such shallow places disease can be a
problem, and fish can also easily die from heat, an algal bloom, or cold.
On the Danube as on the Mississippi, when the pools created by flooding
dry up, people have begun to remove the young fish to save them. In the
floodplain of the Mississippi, an oxbow that in June of 1915 covered

eleven acres was, by November 1915, reduced to a small, marshy pool fifty feet long, thirty-five feet wide, and forty inches deep at the deepest point. All the fish were taken out—more than 150,000 young fish, among which were 30,000 sheathfish, 15,000 crappies, 25,000 sunfish, and 15,000 buffalofish.

During the great flooding of the Elbe that occurred in the rainy year 1926, fish from many ponds moved off. After a while, poisoning was evident in the flooded meadows, probably as a consequence of the putrefaction of the dying grass and the accompanying fermentation of cellulose. Great numbers of fish died in this way and lay on the flattened, rotting grass until workers, some wearing gas masks, hauled them away in wagons and buried them in huge pits.

We have two illustrations from the flooded meadowlands at Dessau [East Germany], plate 7, figures A and B. I thank doctoral candidate Erhard Voigt for these photographs. Remarkably, these assemblages consist almost exclusively of the common pike, ordinarily by no means so common. One picture shows clear evidence of feeding, probably by crows, and a considerable displacement and pulling out of the backbone, while the rest of the carcass has become fairly well mummified. Such severe disturbances in the region of the backbone are usually attributed to feeding by other animals. The same is true for a great many carcasses of fish in the Solnhofen lithographic limestone, as examples in collections at Halle [East Germany], Munich [West Germany], and elsewhere show. The second picture shows an equally badly decayed pike carcass lying in completely rotten meadow grass. The grass itself has been flattened all in one direction by the current and is so neatly arranged that it looks as if it had been combed.

During a swift onset of floodwaters on the Saale [East Germany], Georg Hinsche (69) observed a great many dead and dying toads washed up.

When the rivers in East Africa inundate the wide plains, so many sheathfish wind up in shallow water that they are very easy to catch. During a flood, when the denizens of water, no longer restricted to their normal habitat, spread out temporarily over a wide area, land dwellers, conversely, are severely crowded together. In the floodplain of the rivers of central Germany there is no better means of locating prehistoric settlements than taking a boat during high water to the islands of remaining dry land. In America, the relationship between Indian settlements and rivers is exactly the same. When the floodwaters of the Brazos meet those of the Bernard, Damon Mound [Texas] rises like a peninsula above the surface of the vast flooded area. Since time immemorial there was a settlement there of Carnarvera Indians, members of the Caddo tribe, who conducted a profitable medicine business with a sulfurous earth found in the vicinity. During the [First] World War, this hill, a salt dome with a sulfur cap, was found to contain crude oil. When we read that farmers cannot save their livestock from the floodwaters of the Rio Grande, or when, at home in Germany, herds perish when the coastal dikes rupture, we know these disasters are a consequence of civilization, but forest animals confined to small areas by continuously rising water suffer a similar fate; they either starve to death or drown.

Kormos (93) has drawn on such conditions to explain the masses of bones from the Pliocene at Polgârdi [Hungary]. The area of karstified limestone in question was so small that it could not possibly have supported whole herds of *Hipparion* and gazelle. During the Pliocene, this limestone, honeycomed with caves, rose some sixty-five meters above the plains; it is difficult to see why so many individuals and, in particular, so many species gathered together in the Somlyo Mountains, other than from necessity. Kormos believes the animals were normally not found in the mountains and must have been driven there by either unusually severe flooding or forest or prairie fires. Hunger, even more than predators, could then carry out the work of destruction. Heavy rains would wash the scattered bones into the limestone caves; very small remains, mostly snake vertebra and lizard scales, but also bones from pikas, molelike animals, and other microfauna, usually found their way there by means of the pellets of predatory birds.

Similar events must have contributed to the origin of the *Hipparion* faunal assemblage described by Schlosser (141) from Veles in Macedonia [Greece]. On these bones and in the cementation of the sand around them were observed limestone incrustations whose origin is due to physical processes that occur during decay of muscle, connective tissue, periosteum, and tendons; I was able to make similar observations on recent quicksand carcasses. Just as at Pikermi, in Greece, and Mt. Leberon in the Vaucluse [France], the determining factor at Veles was torrents, or what I prefer to call intermittent streams, which confined the animals to places initially spared by the water, but which overtook them later, either while they were still alive or after they had died from hunger.

Such disastrous cloudbursts and catastrophic flooding happen in modern times, too. Schlosser (141) is probably correct in attributing the accumulation of these mammal remains at Pikermi to a short-term catastrophe. The teeth of a great many *Hipparion* foals and the two-year-olds from *Sus erymanthius* show the same stage of development. The condition of the teeth of individuals of *Ictitherium, Sus erymanthius, Tragocerus,* and *Protragelaphus* enabled Schlosser to conclude that these animals died in the autumn, sometime in October. In Pikermi, the *Hipparian* foals were either fetuses and newborns, or were from ten months to a year old. In Samos, on the other hand, many different age-groups were represented: 0–2 months, 10–12 months, 20 months, 2½ years, and 3½ years. This grouping indicates at least two catastrophes, since only foals of the first three groups could have perished at the same time. One catastrophe seems to have occurred in February or March, the other in October.

Stromer (159) described the interesting vertebrate remains that Erich

Kaiser (84) brought back from Southwest Africa [Namibia]. They were from animals that had lived in a shallow sink with sparse vegetation and intermittent water holes. Several old erosional gullies ran together into this depression. A river suddenly off course submerged the basin, and most of the inhabitants drowned. The gregarious "jackrabbits" fled to a small island where they had to crowd closely together and finally died from hunger or rising water. The floodwaters receded quickly, allowing scavengers to disturb the carcasses and carry off parts of them. Only then were they actually embedded. Consequently, on a stone slab, an ornament of the Munich museum, lie a great number of remains of the "jack-rabbit" *Parapedetes namaquensis* Stromer.

Abel (4) has given us a vivid picture of the way bony remains are concentrated during periodic flooding. At the Vienna Museum's excavations at Pikermi, the bones were embedded in long, narrow lenses in many different horizons of the red clay, the rest of which was completely devoid of bones. Here and there, large blocks of Pentelic marble were found. On the upstream side of one of them lay a tangled mass of bones from many species and individuals. It was first worked from the eastern side, but efforts to separate the bones from each other were in vain. Not until stripping was begun on the west side, in the direction of Megalorheuma Creek, could one bone after another be taken out. It was evident that the individual elements of the bone cluster had been deposited one on top of the other like roof tiles, just as a bank of flat stream pebbles in a river grows upstream. This is what the Americans (see Johnston 83) call an "imbricate structure." At the end of the whole cluster lay the skull of a rhinoceros, which had come to rest crosswise behind a hummock.

In the Middle Sarmatian brackish limestone near Sevastopol [USSR], lenses of osseous breccia are found, which never contain skeletons preserved in entirety. For the most part, the bones are lower jaws and leg bones; *Hipparion* remains predominate. These conditions are described by Borissiak and Schwarz (24, 148). In the Upper Permian on the northern Dvina [USSR], hard limestone concretions with some incomplete and some complete skeletons of considerable size are found in a relatively loosely cemented sandstone, giving the impression that they had been preserved in their natural articulation only because the ligaments were still intact at the time of burial.

11. DEATH DUE TO FLUCTUATION IN SALINITY

The most outstanding example of fluctuation of salinity we can mention is the remarkable desalinization of the estuary of the Bug [USSR], which occurs every year in the spring. Normally, such a phenomenon is caused by meltwater carried by the rivers themselves. However, the floodwaters of the two small rivers that form this estuary, the Bug and the Ingul, have nothing to do with the salinity of the enormous estuary of the Bug, or of the smaller estuary of the Ingul, to the east. Instead, in May, the flood-waters of the Dnieper come in; the enormous volume of meltwater from its large catchment area freshens first the estuary of the Dnieper and then the two others, as Sokolow has described (154). The cloudy fresh water spreads out from the coast inland into the estuary. The quantity of water flowing in varies greatly according to the year. In midsummer, the water starts to become saltier; by fall it has become completely clear and attained its maximum degree of salinity. The invasion by the sea of the Limfjord [Denmark] and the breaking out of the Everglades into the Gulf of Mexico are further examples of this phenomenon, which we explore elsewhere.

12. DEATH WHEN BODIES OF WATER DRY UP

In arid and semiarid climates, the existence of bodies of fresh water is by no means as assured as it usually is for us who live in more humid zones, and even here noticeable fluctuations in water levels are not unusual. On prairies, there are countless lakes; in winter, they are inhabited by crayfish and visited by countless shorebirds, but soon thereafter they dry out, leaving nothing but hard, barren ground. Some creatures, like the turtle, the crayfish with its tunnels, and even the muskrat (whose subterranean burrows, which severely damage the banks, follow the fluctuation of the water level so that a little water always remains in them), are suited to the abrupt change. Those that are not are wiped out. The same mode of death often occurs on a broad scale when rivers flood: fry are driven into closed pools where at first food conditions are exceptionally favorable to growth, but from which, if a drought occurs, there is no escape. Irrigation produces the same effect over wide areas. A considerable volume of river water is pumped continuously into the large rice canals and carried to the rice fields, where it evaporates without a trace. During the process, many young fish end up in the canal system. At the end of the period of irrigation, no more water is added, and the canals dry up; the beds become as hard as stone and myriad dessication cracks appear. The tough needle gars (*Lepisosteus osseus*) hold out for a long time in the last bit of muddy water. They can do so because they have an air bladder, into which they gulp air, and other adaptations. The fanning of their tails and fins is so strong that bowl-shaped depressions are formed in the mud at the bottom of the deepest places where they gather, and the carcasses finally come to lie side by side, alone or accompanied, at the most, by a few catfish and bowfins (*Amia calva*). The flesh decomposes quickly or is eaten by small animals, and the scaly skin encloses only empty space. Look at the perforated specimen in plate 8, figure A, observed in the old rice canal at Bay City [Texas] on July 5, 1925. In a pothole, also fanned out by the fish as they clung to life, we see a bowfin and two gars (pl. 8, fig. B).

Exactly the same thing occurs under natural conditions, as for example, in the oxbows of the Colorado, which dry up in the summer after the floodwaters recede. A fine, red-violet mud breaks up into a mosaic of drought cracks, whose sharp edges are softened by rain. In the muddy clay there are sticky, pale yellow balls of alligator excrement, often in a perpendicular spiral, just as they were excreted. Behind one oxbow there was a large alligator den, and in a few places there were remains of turtles and fish. We often see such things in the wooded floodplain around Bay City.

Concentrations of animal carcasses due to drying up of the habitat are doubtless more often seen in the geologic past. Wepfer (173) was correct in rejecting the assumption that the *Mastodonsaurus* in the Buntsandstein of the Black Forest near Kappel in Baden [West Germany] had lived burrowed in the mud at the bottom of a pool, had died when it dried up, and had been embedded on the spot. There are other reasons for this concentration of remains. Just as with the alligators of Smithers Lake, many skulls (twenty-five out of thirty-five) are lying upside down. Not one complete skeleton is present. On the other hand, the original articulation is not completely absent: when the water dried up, part of the carcasses were still held together by tendons. Wepfer writes:

> Vertebrae of the forward end of the body are lying immediately behind a skull; several tail vertebrae are lying together; numerous ribs, certainly belonging originally to the same individual, are lying together; here and there, ribs and vertebrae are especially numerous and the only bones present; several superciliary arches are lying near the vertebrae; the middle gular plate lies more often near the lateral one; one left lateral gular plate lies by the left rear corner of the skull, and the right one, only a few decimeters away. Now and then, lower jaws lie right next to the skulls to which they obviously belong, remarkably often in the same position—crossways in front of it. . . . However, the lower jaws are always (with one exception) separated from the skulls, and the two rami have come apart at the forward symphysis; indeed, it seems that a certain sorting of the bones has taken place: lying on one sandstone slab measuring exactly one meter square are four rami. Therefore, before the final burial, some rearrangement must have occurred, probably due to the motion of the water (or due to other animals? No tooth marks are seen).

Abrasion from being tumbled about in water is nowhere observed.

Compared to skulls, pelvic and thoracic girdles and extremities are surprisingly rare; the disproportion is, however, not as great as at Bernburg [East Germany]. The extremities are the most easily detached part of the carcass, and probably most of them were already lost. Apparently the concentration of bones was arranged in a sort of fringe, just as was the unusual deposit of *Trematosaurus* and *Capitosaurus,* which were found with fish, the phytosaur described by Jaekel, remains of *Trachelosaurus,* and abundant *Pleuromeia.* It is important to make sketches of the relative positions of the remains on such a typical depositional plane, and the opportunity to do so at future excavations should not be missed. At Bernburg, there are no extremities at all. Perhaps the animals drifted around for so long that the legs decayed or were eaten. It is also possible, however, that some animals, especially the larger individuals, came to rest in shallow water intact and then decayed, and that the lighter bones thus set free were carried farther toward the shore. At Kappel, as at Bernburg, the concentration of remains is so large that it evidently accumulated after death. The rich find at Kappel covers an area of forty-five square meters, and the slope ratio of the superpositional plane shows that the area once surrounded a body of water.

During the dry season, the concentration of creatures in ever-shrinking bodies of water is fatal for many animals. The vast numbers of hippopotamuses in African rivers, about which much has been reported, often appear even larger when the creatures are compelled to crowd together during the dry season. Rivers are reduced to stagnant branches and pools, each surrounded by a wide band of swamp beyond which the earth is crisscrossed with drought cracks and trails. When that happens, black hunters can easily encircle the small ponds, and the large animals fall victim to their spears.[11] The same thing happens to crocodiles and large catfish. Waibel (168) writes as follows:

> As the lives of plants and animals in the grassland have a pronounced periodicity, so do the lives of the people who live there. The rainless dry season is for them, too, a time of scarcity. Plants wither; the earth dries out. If the grassland dweller does not want to starve, he must gather supplies for this lean, unproductive time. The fruits of the forest, which rot and spoil easily, are little suited to preservation; on the savannah, as in all grasslands of the earth, they are replaced by durable seeds. Grasses, especially different kinds of millet, are the main nourishment of the people of the grassland. In addition to seeds, they also gather large quantities of fish. At the end of the dry season, when the many rivers carry little water and the small streams are completely dried up, men and women go out to catch fish, mostly catfish. Great quantities of these are dried and preserved. Next to meat, they are a favorite supplement to the national dish of the grasslander—*fufu.*

A similar situation occurs even in our part of the world. Among catfish, the care of the brood is confided to the male, who watches over the eggs

11. In exactly the same way, the great winter fish catch on the Danube floodplain in no way represents the product of the waters where the fish were caught. The fish actually matured in the flooded areas. The deep water of permanent lakes is only the refuge where they gather to spend the winter. There, they do not gain weight but lose it.

and aerates them with movements of his tail. This behavior contributes greatly to reproductive success but can be a disaster for the parent fish. When floodwaters retreat, the male often stays behind too long and is then easy to catch. The rivers of the United States are extremely rich in catfish species. Today, the catch is less than it used to be, but a few years ago it reached fourteen million pounds. The blue catfish of the Mississippi weighs up to 125 pounds. Like many members of its species, it can live in very dirty, murky, oxygen-poor water. It is one of those fishes that cling obstinately to life, that stay alive the longest during a drought, and that eat the other fish they find crowded in with them in the small bodies of water. It is the same with the gar. I have seen many desiccation depressions in which the dried-out carcasses of only these two fishes, or of only one of the two, lay on the bottom. In 1912, during excavations in the town of Nottingham [England], Swinnerton (161) found, in a fine-grained sandstone of the Lower Keuper, numerous specimens of *Semionotus* and of the new genus *Woodharpea*. Thin films of clay, along which the sandstone split, formed smooth or undulating planes. The fish carcasses lay close together, 0.3 to 0.6 centimeters beneath such a plane. The person who directed the excavation came to the conclusion that at the beginning of a drought, the animals buried themselves in the sandy mud and died before the water returned. If the overlying bedding plane was smooth, then the fish were not so deeply buried, but only partially covered with sediment, so that only the head was sheathed thickly in mud.

As a body of water dries up, the creatures that live in it move toward the center of the basin along with the retreating water. Bivalves, especially, can come together in huge masses on a small surface.[12] Tench and Prussian carp burrow deep into the mud crust and survive the dry period in a kind of torpor. The European mudeater buries itself as much as a meter deep in mud that later becomes completely hard. I myself have observed that small turtles get stuck in mud that hardens too quickly. In one example that I photographed, although the turtle was caught fast, drought cracks spreading out from his body in every direction had freed the left foreleg and the head, and traces of the animal's struggle were etched in the mud. According to Semon's description (151), the Australian *Neoceratodus forsteri* are just like catfish and billfish. Their spawning time falls right at the end of the driest period, which lasts from the end of August until the middle of October, when water levels are at their lowest. Thus, the young are in no danger of being damaged by further drought. During the dry season, the fish remain in depressions in the riverbed, which is usually wide and deeply incised, but almost completely devoid of water. Once or twice a year, torrential rains fall and the channel fills. Sometimes even this does not happen and the water ceases to flow at all, but the water holes

12. *Dreissenia polymorpha,* observed by Antipa in the area of the Danube, attaches itself to retreating freshwater clams and is carried back toward the depths of the permanent lake.

themselves never dry up. Perhaps the survival of *Neoceratodus* is attributable only to this phenomenon. Today the fish is found only in the Burnett and the Mary, and not in the rivers to the north and south of them. Post-Pliocene finds prove that the other rivers once contained these fish too, and not so very long ago.

The complete drying-up of a river system is sufficient to exterminate its fish. It has never been observed that *Neoceratodus* buries itself when the water dries up. If the water holes dry up completely, then eels and other fish burrow deep in the mud, but *Neoceratodus* dies, although not until the very last; when many fish have died as a consequence of the water's going bad, *Neoceratodus* is still fresh and lively, which is probably the main biological significance of its lung breathing. Quite a number of *Neoceratodus* remains from the Triassic can probably be attributed to drying up of the habitat. Vollrath (165) is of the opinion that the localities in the Stubensandstein south of Phohren near Donaueschingen [West Germany] also represent the creature's habitat.

Whereas *Neoceratodus* has only a single "lung" and rises regularly to the surface to gulp air, the gills of *Lepidosiren* are located far back on the body, but the lung is paired. When *Protopterus,* the African lungfish, cannot breathe air, it dies after two to three hours, but out of water it can remain alive for more than twenty-four hours. The Gambia and many other rivers of equatorial Africa are its native habitat. During the rainy season, it goes out over the flooded, swampy areas to feed, mainly on frogs and crayfish. When the water dries up, the fish does not return but buries itself one-half meter deep in the mud and protects itself with a mucuous-lined capsule. In the area around Lake Charles, Louisiana, at the edges of flooded areas, I often came upon buried specimens of the eel-like *Amphiuma means* Gard., which the inhabitants call "poison eel" in spite of its four, albeit very short, legs.

Batrachian larvae present an interesting example of the concentrating effect of drying up (pl. 8, fig. C), analogous, in a way, to the fanned-out depressions made by fish when water recedes. I took the picture April 18, 1925, at the rapids on the Brazos River between Rosenberg and Richmond [Texas], when the water was very low. At that point, the river, feared because of the way it changes its course, had recently cut through the calcareous sandstone of the Reynosa Formation. There was still a shallow puddle on the bank, but as I took the picture, most of the tadpoles darted away, leaving only a few behind. Notice that the entire bottom of the puddle is honeycombed with contiguous, hexagonal depressions of even size, which correspond to the tadpoles' radius of movement. When the water dries up completely, these depressions are usually lost in the drought cracks, but that does not mean they would never be preserved as fossils. People have often thought they have found such remains, but everything that until now has been described as such is, according to Abel (7), a system of depressions that often occurs when two ripple marks intersect at right angles. Pia (127) asks whether problematic structures like *Palaeodictyon* from the Eocene sandstone of the Tropfberg near Vienna [Austria], described by Theodor Fuchs (56), cannot be attributed to such evidence of life. We need not think only of batrachian larvae, because many different organisms are capable of forming similar structures. Tadpoles have also been found on completely smooth bedding planes where there is no evidence of drying out.

Such "tadpole nests" will probably be found as fossils one day, but what E. Hitchcock (70) described in 1856 as fossil tadpole nests in the New Red Sandstone from the north bank of the Connecticut River, wrote up as *Batrachioides nidificans,* and showed a picture of are, according to Shepard (152), hollows formed by intersecting ripple marks. Their outlines are mostly rectangular, or at least close to it, and there is too much variation in their size.

13. OVERCROWDING OF ANIMALS DURING DROUGHT

It is well known that on the African steppe game used to be abundant. To help us visualize just how abundant, I refer to two eyewitness reports of vast herds of springbok: the first, from a book by Leo Waibel (168), is an account by Le Vaillant written in 1795; the second, by Eberhard Kretschmar, was written in 1853. The Frenchman said he had encountered a herd of from sixty thousand to eighty thousand of those animals, which filled the whole of a small valley near the Orange River [Namibia and South Africa]. Kretschmar, a German doctor, found himself on a lonely farm on the border of the Karroo [South Africa] when a large herd of springbok signaled its approach:

> Early one morning I positioned myself with a group of hunters at a narrow pass, with mountains extending long distances to the right and left. The herd had to come this way. A muffled rumble preceded it, like surf on a distant ocean or the rush of wind from an oncoming storm. The hunters quickly climbed up on high piles of rock to be out of the way of the tide of animals. Now the springbok came on—by twos, by threes, by tens, twenties, in larger and larger troops; indeed, there came to be as many as three or four hundred in each separate cluster. Finally an unbroken, brown and white mass poured through the pass, which was at least eight hundred paces wide, not like separate animals but like a river. Without interruption, the living flood rolled by. The dogs had long since disappeared; when the first springbok had appeared, they had mingled with the herd and were carried off by waves of onrushing animals. A native was also caught up in the maelstrom. He tried in vain to escape, struggled for a moment against the current like a drowning man and then sank, and there was nothing but a chaos of dust and springbok. Once again his

head bobbed up, only to instantly disappear again. The mighty throng flowed on. The Boers who were with us, and who were quite accustomed to huge herds of cattle, estimated there were twenty-five thousand springbok in this migration.

Such a vast number of animals is observed only during the dry season, and it is a climate-induced, seasonal concentration drawn from a wide area.[13] In southern Namaland [Namibia and South Africa], between the lower Fish and the Orange rivers, there was, according to Waibel, another herd numbering fifty thousand. During the rainy season, the animals dispersed around the isolated pans of the southern Kalahari [Botswana and Namibia]. In July or August, when the first winter rains fell in the area south of the mighty Karras Mountains, they gathered together and mi-

13. Namaqua grouse (*Pteroclurus namaqua*) gather in flocks of as many as sixty thousand to fly in search of water, and when they find it, they overcrowd the water hole. Otherwise, they live dispersed and isolated within their endless habitat. Hesse (animal geographer) describes the same situation:

> Moreover, mammals gather around water holes, permanent springs, puddles that have not yet dried up, remains of ponds and lakes, around the pans in South Africa, and many even around brackish water. In the rainless spring of 1887, A. Walter found the *Antilope subgutturosa* [*Antilope cervicapra*] in countless numbers around the spring Adam-ilen on the Afganistan border. The more infrequent the water holes, the larger the gathering of animals. Some of them stay in the area close to the spring, while others spread out a bit more. Monkeys go no farther than 4 to 6 kilometers from water; likewise, rhinoceros, waterbuck (*Kobus*), and reed buck (*Cervicapra*) [*Antilope cervicapra*] remain close by; in the Australian bush, the small finch *Taeniopygia castanotis* is an indication that water is near. The elephant, on the other hand, will go as far as 30 kilometers and more from water, and the Namaqua grouse can be as far as 175 kilometers from it. At such a water hole, then, there is an astonishing traffic—an uninterrupted coming and going of animals seeking to quench their thirst. During the day and at dusk, the visitors are mainly birds, many of which, as in South Africa, are grouse, doves, and quail, in flocks so huge they darken the sun; mammals— herbivores and carnivores—prefer to come at night, singly, in groups, and in herds. Even basins where the springs flow heavily can be drunk completely dry during a single night.

> When, however, the springs on the savannahs run dry and the last pools evaporate, times are hard for the mammals; they have to migrate in search of new habitats that offer more favorable conditions. Domestic animals are in the same situation, and their owners, nomadic shepherds, must follow them. The migrations of savannah dwellers such as zebras, antelopes, and ostrichs usually follow a regular pattern, and their predators—lions, leopards, and hyenas—follow them. Famous for their mighty herds in migration are the springboks (*Antidorcas marsupialis*) of South Africa, whose number has been estimated at over forty thousand head, and even as high as sixty or eighty thousand. Sometimes, however, in especially dry years, migrations at unaccustomed times lead to catastrophe; at such times, as in the winter of 1863 in South Africa, animals (*Cephalophus mergens*) [duikers; *mergens* no longer a valid species name] driven by hunger and thirst approach human settlements and even enter villages, only to die there by the thousands.

grated west. Animals in the east and south African steppe gathered in huge numbers only during the dry season, when they were forced to remain near permanent springs. In the rainy season they paired off and spread out over a much wider area, and the apparent superabundance disappeared. Waibel described vividly the dependence of game upon water holes. It is these areas of the steppe that are so completely denuded by the animals. According to Passarge, the holes, basins, and depressions they make drinking, bathing, scratching, and rooting around are then enlarged by wind and water to form blowouts.

Besser (23), too, observed in Africa that, in spite of statements to the contrary, game kept to a particular trail, albeit one that shifted considerably as water and food conditions changed during the course of the year. However, at a particular time, the game always went back to its familiar shelter and grazing land. Food plays the ultimate role in the migrations of animals, whether they be squirrels, rats, lemmings, wapiti, buffalo, or springbok. It is the same with the African donkey, migratory birds, and even some reptiles, although the demands reproduction places on reptiles, and on fish, too, may seem to be the main cause of migration.

Just as migration in search of necessary food or life-sustaining water introduces an unpredictable element into the lives of mammals, so the migrations necessary for reproduction for amphibians, many reptiles, and fish, from the ocean to fresh water or the reverse, are dangerous, but are nevertheless carried out unswervingly. During every such migration there are accidents, mainly caused by overtiring or complete exhaustion. Mosso has done research on quail that migrate across the Mediterranean Sea from Cap Bon [Tunisia] across Sicily to Rome, a distance of 449 kilometers in nine hours, and found that when they arrive, their brains are completely bloodless. Consequently, while flying at full speed, they may sometimes crash into houses and trees.

We are more and more accustomed to thinking in terms of an average climate, determined statistically by measurements taken over a period of many years. This oversimplification is misleading and causes us to become careless. Even here in Germany, people have planted grain and root crops in many floodplain meadows, although the peasants knew well enough why these fields used to produce only hay. The result is sometimes severe flood damage. Other weather conditions not corresponding to the normal patterns for a year can occur: lethal drops in temperature; extremely heavy rainfall; unusually severe storms; snow lying too long on the ground; absence of rain for too long a time. Such events often presage a gradual change in climate, and short-lived humans do not take them properly into account. I must quote here Darwin's classic description (37) of the disastrous effects of drought in South America, although it may be familiar to many readers.

While travelling through the country, I received several vivid descriptions of the effects of a late great drought; and the account of this may throw some light on the cases where vast numbers of animals of all kinds have been embedded together. The period included between the years 1827 and 1830 is called the "gran seco," or great drought. During this time so little rain fell, that the vegetation, even to the thistles, failed; the brooks were dried up, and the whole country assumed the appearance of a dusty high road. This was especially the case in the northern part of the province of Buenos Ayres and the southern part of St. Fe. Very great numbers of birds, wild animals, cattle, and horses perished from the want of food and water. A man told me that the deer* used to come into his courtyard to the well, which he had been obliged to dig to supply his own family with water; and that the partridges had hardly strength to fly away when pursued. The lowest estimation of the loss of cattle in the province of Buenos Ayres alone, was taken at one million head. A proprietor at San Pedro had previously to these years 20,000 cattle; at the end not one remained. San Pedro is situated in the middle of the finest country; and even now abounds again with animals; yet, during the latter part of the "gran seco," live cattle were brought in vessels for the consumption of the inhabitants. The animals roamed from their estancias, and, wandering far southward, were mingled together in such multitudes, that a government commission was sent from Buenos Ayres to settle the disputes of the owners. Sir Woodbine Parish informed me of another and very curious source of dispute; the ground being so long dry, such quantities of dust were blown about, that in this open country the landmarks became obliterated, and people could not tell the limits of their estates.

I was informed by an eye-witness that the cattle in herds of thousands rushed into the Parana, and being exhausted by hunger they were unable to crawl up the muddy banks, and thus were drowned. The arm of the river which runs by San Pedro was so full of putrid carcasses, that the master of a vessel told me that the smell rendered it quite impassable. Without doubt several hundred thou-

---

*In Capt. Owen's Surveying Voyage (vol. 2, p. 274) there is a curious account of the effects of a drought on the elephants, at Benguela (west coast of Africa). "A number of these animals had some time since entered the town, in a body, to possess themselves of the wells, not being able to procure any water in the country. The inhabitants mustered, when a desperate conflict ensued, which terminated in the ultimate discomfiture of the invaders, but not until they had killed one man, and wounded several others." The town is said to have a population of nearly three thousand! Dr. Malcolmson informed me that during a great drought in India the wild animals entered the tents of some troops at Ellore and that a hare drank out of a vessel held by the adjutant of the regiment. [Quoted from Charles Darwin, *Journal of Researches into the Geology and Natural History of the Various Countries Visited during the Voyage of H.M.S. Beagle round the World* (London: M.M. Dent; New York: E.P. Dutton, 1906), pp. 125–27]

and animals thus perished in the river: their bodies when putrid were seen floating down the stream; and many in all probability were deposited in the estuary of the Plata. All the small rivers became highly saline, and this caused the death of vast numbers in particular spots; for when an animal drinks of such water it does not recover. Azara (*Travels,* vol. 1, p. 374) decribes the fury of the wild horses on a similar occasion, rushing into the marshes, those which arrived first being overwhelmed and crushed by those which followed. He adds that more than once he has seen the carcasses of upwards of a thousand wild horses thus destroyed. I noticed that the smaller streams in the Pampas were paved with a breccia of bones, but this probably is the effect of a gradual increase, rather than of the destruction at any one period. Subsequently to the drought of 1827 to '32, a very rainy season followed, which caused great floods. Hence it is almost certain that some thousands of the skeletons were buried by the deposits of the very next year. What would be the opinion of a geologist, viewing such an enormous collection of bones, of all kinds of animals and of all ages, thus embedded in one thick earthy mass: Would he not attribute it to a flood having swept over the surface of the land, rather than to the common order of things?

(I experienced the same thing with armadillos—*Tatusia novemcincta* [*Dasypus novemcinctus*]—in the most southerly part of Texas.)

It is too bad we have so few contemporary reports of analogous events, which happen much more often than seems to be indicated by accounts handed down to us. We need think only of the population movements of central Asia; one wonders how many separate meteorological events must have been responsible for them.

In 1925, an unusually severe drought afflicted the southern part of central Texas. When I went to Taylor, Texas, practically no rain had fallen for eleven months, and I still remember the desolate appearance of the countryside. The fields, already planted in vain for the second time, were completely burnt up, and everywhere, strange mirages formed in the layer of hot air above the earth. An exceptionally large number of livestock perished in the drought, and in a few areas the lack of feed was so severe that live cattle were put up for sale at only two or three dollars a head, and even at that price hardly a buyer was to be found.

The intermittent growth of the horns caused by such unhealthy feeding situations is curious. In bivalves and turtles, we can see every pronounced drying-up of the environment as an interruption in the growth of the exoskeleton; we can read the fate of cattle, too, in the growth of their horns. In wet years, when grass is abundant, strong, even growth takes place. When drought sets in, with its lack of food, stepped, uneven growth occurs. This was typical of the cattle that died during the great norther at the end of 1924; they had already endured drought and lack of food. We see the same thing in humans who have been severely ill during the period of their second dentition: zonal growth disturbance can be observed on the front teeth.[14]

An important point concerning the preservation of remains is that normally during drought and famine, animals temporarily frequent dried-up ground that later often becomes an area of sedimentation.

In African lakes, crocodiles, after reproduction has taken place, hold out temporarily in the band of vegetation along the shore because the evaporating body of water has retreated by several kilometers, and the muddy pools remaining cannot provide a suitable habitat for so many individuals. There are also, however, examples of extreme, long-lasting drought driving crocodiles together in masses. Wiman reports (176) that according to Gadow, all the swamps and lakes on Marajo Island, in the mouth of the Amazon [Brazil], once dried up during a severe drought. Thereupon the alligators tried to migrate to the next river but often paid for the search with their lives. In one place, eight thousand carcasses were found lying together and at the end of Lake Arary, more than four thousand. Wiman compared this mass death caused by drought with the "nests" of *Belodon* in the Keuper of Württemberg [West Germany].

14. DEATH DUE TO HUNTING

Soergel (153), to whose work we owe so many valuable ideas, took an extraordinarily important step when he investigated the Pleistocene mammalian fauna to see if human hunting practices had affected its makeup. He considered the game's viability and capability of self-defense, the contemporary cultural level of the humans, and—what interests us most here—the makeup of the fossil material. At Taubach-Ehringsdorf [East Germany], only 16 percent of the *Elephas antiquus* found were adult animals with serviceable third molars. In geologically older Süssenborn [East Germany], on the contrary, 78 percent of the animals were adult; at Mosbach [West Germany], 61.5 percent of the *Elephas antiquus* and 58.3 percent of the *Elephas trogontherii [Mammuthus trogontherii]* were adult; and at Mauer [West Germany], the same high percentage of *Elephas anti-*

14. Not only severe disturbances in general nutrition but also evidence of childhood illnesses are discernible for a long time in transverse ridges in tooth enamel, and we observe the same phenomenon in human fingernails until they grow out. I have also observed uneven, stepped growth in bulls' horns. B.G. Gruber, head of the Institute of Pathological Anatomy at the University of Innsbruck, was, without doubt, completely correct when he told me in a letter that the regular ridges in the horns of female cattle are caused by pregnancy and lactation, and that people use these growth ridges to tell the age of milk cows. The growth of human fingernails and toenails can also be influenced by the same processes.

quus calves was found as in Taubach. Soergel related the high number of young elephants in Mauer, Taubach, and an English locality to hunting during Paleolithic times. The Predmost mammoths [ČSSR] seemed to have died in a catastrophic way; Soergel thought the herd was annihilated by an epidemic, and he called the occurrences at Cannstadt [West Germany] and Emmendingen [West Germany] dying grounds. An inquiry into the single mammoth find in central Europe revealed most of the animals to be older adults. In the Baden loess [West Germany], 87 percent were mature animals, and in the lower terrace lying on top if it, 65.7 percent; these figures by no means indicate intensive hunting. Research on horses and species of rhinoceros has yielded similar results. At Mauer, an unusually low number of rhinoceros calves were found, while at Taubach there were many. Among the cattle, 21.1 percent of the animals in Mauer were young, while at Taubach the figure is 57.8 percent.

In the literature on Paleolithic hunting, the long-known mass of horse bones from the Aurignacian in the Saône Valley [France] has always been an important element. This horse "magma," a homogeneous layer twenty to one hundred centimeters thick, covers an area of more than one hectare and is well buried in some places and exposed in others. Over one hundred thousand individuals are represented in this carcass assemblage. Because it is found at the foot of a cliff, the explanation was given that primitive hunters surrounded the whole herd and set fire to the plains, forcing the animals to plunge over the cliff to their deaths. But the horse skeletons were by no means complete; for the most part, only feet and spinal columns were found, and other skeletal elements were missing almost entirely. This assemblage probably represents the accumulation of kitchen rubbish over many generations and not a single great hunt. Meanwhile, beneath the horse layer at Solutré [France], human skeletons have been found.

One primitive hunting practice, that of driving game onto unstable ground where it sinks in helplessly and is then easier to catch, seems to have been widespread. The bone assemblage of giant elk found in the Irish moors around Dublin may have come about in this way. North American Indians used this practice in hunting elk, and the aboriginal hunters of the Dutch East Indies drove elephants into swamps. Carnivores hunting in packs do exactly the same thing, and in the asphalt swamp at Rancho la Brea, victims of wolves' game drives were found. The moas of New Zealand met the same fate. It makes no difference whether humans or predatory animals chased the creatures into the swamp—the result is the same. In this sense, we can use the concept of hunting in the geologic past, too. The muddy swamps of South Africa are a good place to look for ivory from dead animals. It is difficult to say whether such sunken carcasses are due to normal accidents or to pursuit by hunters. Klähn (91) has noted that losses to hunted herds due to sinking into a swamp occur mostly among the young and the very old.

English authors such as Keith (87) would like to attribute the dying out of pygmy elephants and hippopotamuses on Malta to Neanderthal man. The biggest cave on Malta is from 20 to 60 feet wide and 18 feet high. The deposits on its floor are from 16 to 18 feet thick. Beneath them is a clay devoid of fossils. Above them lies a layer 1.5 to 2.5 feet thick composed of water-tumbled bone fragments of three extinct species of elephant and two of hippopotamus. Thousands and thousands of animals must have come into this cave for such a deposit to have been formed. We might well ask whether this mass of bones was not carried into the cave by the surf. Keith's comments, however, are accompanied by a drawing of a fanciful scene showing a whole herd of elephant and hippopotamus being driven into the cave by Neanderthal hunters armed with clubs and stones.

It is certainly an old hunting trick to cause a herd to panic and stampede.[15] Primitive men could have learned this by observing the hunting tactics of predatory animals. In the course of such a stampede, animals would always have accidents, break bones, fall off of high places, get stuck, exhaust themselves, and so on. African aborigines understood that by setting the grasslands afire, they could throw herds of antelope into such confusion that they could not break away and were easy to slaughter. It is reported that during an autumn storm on the German coast in October 1926, a herd of sheep was caught in the flood tide on Padelaks-Hallig near Husum [West Germany]. Some of the animals plunged into the open sea, and one hundred sheep drowned. Darwin (37) reported earlier that herds of cattle on the pampas plunged into the Parana River, blocking the branch of the river near San Pedro with twenty thousand rotting carcasses, just as I saw in the Intracoastal Canal, south of Bay City near Sargent. He went on to describe how panic-stricken horses plunged into swamps, where more than a thousand carcasses accumulated. Wiman (176) reported that in Lapland, reindeer pursued by wolves or fleeing a snowstorm plunged over steep cliffs to their death or dashed around all day and drowned in lakes. According to Abel (4), Lloyd Dawkins described how packs of hyenas caused frightened animals to leap from precipices, and in recent American hunting magazines it has been reported repeatedly that in the spring, carcasses of does driven over cliffs by wolves were found.

Not infrequently the predator, in hot pursuit, meets its own death. When an old, steep-walled, completely inaccessible quarry near Probstzella was reentered through a tunnel, the carcass of a roe deer was found, as well as that of the dog that had been chasing it. Such observations have also been

---

15. We read over and over how easily herds of domestic animals can be driven to panic. A newspaper reported the following incident: "Panic in a herd of sheep. Near Neuengeseke, near Soest, Westphalia [West Germany], a stray dog got into a herd of sheep during the night and mauled several. The animals panicked, and a total of 104 were killed." Exactly the same thing happens among humans, as we see to our sorrow whenever fire breaks out in a theater or school.

made in mines. Plate 9, figure A shows a weasel that had entered a mine tunnel near Eisleben [East Germany] and lost its way while chasing a mouse. It drowned in the drainage ditch and remained lying on the gravel, bits of which adhere to the back of the piece. Dripping water made a hole in the carcass, creating a round pit in which lies a piece of rock encrusted with limestone. The ribs and pectoral girdle are correspondingly somewhat displaced. Remains of the encrusted skin are seen lying in folds around the bones, starting at the skull and extending on all sides. Sintered hair is also found around the hole. The right scapula is visible at the lower end of the piece. This specimen is in the private collection of a teacher named Bessler, who lives in Eisleben. The carcass of a young cat found in an old wall is perhaps due to similar circumstances; after being mummified, it was impregnated with limestone. This specimen is in the Krahuletz Museum in Eggenburg [Austria] and is shown in an illustration by Schaffer (138, p. 413, fig. 427).

Although in prehistoric times only the extinction of the bear was directly influenced by humans, the situation in modern times is different. The number of creatures that have been hunted to extinction or near extinction by humans is sizable. In Labrador, the caribou has almost disappeared due to reckless hunting, and wolves and Indians dependent on caribou for their livelihood are affected by this loss. (We almost always see such indirect consequences when a closed biotic community is attacked. Rinderpest in East Africa provides another example. In many parts of the land, it annihilated not only cattle and buffalo, but caused many lions and other predators to starve and brought the Masai to the brink of ruin.) Thienemann (163) has explored this problem more extensively than anyone else. The great condorlike vulture *Gymnogyps californianus* is richly represented among the Pleistocene fauna of La Brea. When white men first came to California, deer, cougar, and these huge condors formed a closed biotic community. The first California hunters had all they could do to protect the game they had shot and hung out or stored from them. But the populations of deer, elk, antelope, and mountain sheep decreased rapidly, and with them went the cougar, who preyed on them. It was the carcasses of these animals that the condor fed upon. The settlers protected their introduced domestic animals with rifle and poison, with the consequence that today the California condor is one of the rarest of birds.

There are other examples of extermination caused by humans. In 1893–94, womens' muffs made from the skins of colobus monkeys were fashionable. The creatures were shot in unbelievable numbers and soon became rare. Then there are the bird of paradise, the egret, and finally the bison, which came to be hunted only for its hide; the so-called sport of hunting also bears much of the blame for the bison's disappearance. The quagga was exterminated in the Cape Colony in 1870, further hunted in the Orange Free State [South Africa], and ten years later eliminated there, too, by the Boers. They shot the animals to make grain sacks out of the hide of the upper hind leg. Burchell's zebra was a victim of the Boer Wars.

Animals have also been eliminated for gastronomic reasons. The prairie chicken (*Tympanuchus americanus americanus*) [*Tympanuchus cupido*], so numerous even recently, will soon be completely extinct. I saw it in the most isolated parts of West Texas, and even there, only rarely. It used to be present in vast numbers over the whole Mississippi Basin from Manitoba south to Louisiana and Texas and west to Colorado. The remaining birds are now being mercilessly picked off—all because of their tasty flesh. The gastronomic issue is often decisive. The existence of the hippopotamus in parts of Africa such as Rufidji, where the indigenous people despise its flesh, is hardly endangered. Without firearms, it is not worth the risk to hunt the animal only for its hide and teeth; in other areas, however, destruction is assuming large proportions.

Within recent decades, a parallel to the extermination of the giant turtles of Réunion [Mascarene Islands, Indian Ocean] has taken place: the relentless killing at sea of the giant Galápagos tortoises of the Pacific. Beebe (20) has recently assembled all the reports on these giant tortoises, known since 1684. In all of the accounts, the delicious taste of the flesh was emphasized. Dampier (Beebe 20) wrote as follows: "I heard that very big ones can also be found on the island of St. Lorenz or Madagascar, and in the English woods of an island nearby, called either don Mascarehna or Bourbon, belonging now to the French; but whether they are as big, fat, and tasty as these, I do not know." In 1685, Captain Davis took sixty crocks of turtle oil along with him as a substitute for butter. Turtles were used to provide fresh meat for the ships' crews, to protect them from scurvy. The creatures were said to have been able to live for a quite a while without eating. In 1813, Captain Porter took fourteen tons of turtle on board in four days. Darwin's descriptions are well known, and the best of the most recent reports comes from Beck in 1905. He spoke of the speed with which the animals were killed. The workers killed many animals for only a few pounds of meat and some fat to go along with it and let the rest rot. Now that we are beginning to extract the oil, annihilation of the turtles is proceeding rapidly. Today, only a few of the fifteen species remain. Just look at Beebe's picture of the large shells of hundreds of slaughtered turtles on the island of Albemarle [Isabella Island, Galápagos]. They lie there piled up as though they had died naturally from drought. The Spanish giant tortoise, exhibited today in the museum in Madrid, also lay in such remarkable piles, which are found here and there throughout the Spanish Miocene. In one place, twenty-one of these *Testudo oliviari* were found close together.

To historic incidences of extermination belong the often recounted examples of animals that find themselves in restricted habitats under very unstable conditions. I cite here the huge birds of New Zealand; Steller's sea-cow (*Rhytina stelleri*) on islands in the Bering Sea; the two giant pigeons—*Didus ineptus* of Madagascar and *Didus apterornis* of Bourbon; the *Pezophaps solitaria* of Rodriguez; the red chicken, *Aphanapteryx;* and the giant rail, *Gallinula gigantea.* Only a short time after the discovery of the Mascarene Islands in the Indian Ocean, the aforementioned birds were completely exterminated from the islands of Mauritius, Bourbon, and Rodriguez. The same fate met the giant auk, which has been gone now for more than sixty years.[16]

However, in all of these instances, as Walther (171) has emphasized, the population was isolated, a relict, the individuals crowded into a restricted habitat—in short, a situation conducive to eradication. Walther (171) described how disastrous it is to force certain species from their customary homes and mentioned that the Pyrenean mountain goat does not reproduce in Switzerland, where it was introduced, because in its new home the snow does not melt until a few weeks after the young are born.

An example taking place in our own time shows emphatically that an attack on a way of life is often much more disastrous than a murderous hunt. I am thinking of the huge herds of wapiti in North America, which are doomed if vigorous help is not forthcoming. Originally there were about fity thousand individuals in the hilly part of western Wyoming. In summer, the herd dispersed throughout the high mountains, where grass was plentiful. When heavy snow covered the peaks, it decended to the lower, windswept ridges, where the necessary food was laid bare. Thousands of wapiti spent the winter on these exposed ridges, the last to be covered with snow in winter, the first to be free again in the spring. The Snake River flows through Jackson Hole, the herd's habitat. Along its course are fairly extensive damp meadows, where natural grass hay grows abundantly. When heavy snows finally drove the huge herd into the valley, it could still graze the long meadow grass and find food here and there along the sides of the valley. Then settlers came. Farmers claimed the meadows of the valley floor; they herded their cattle into the hills for spring pasture. When snow forced the cattle to the floor of the valley, they grazed it right down to the ground, leaving nothing for the wapiti. The farmers had already mowed the valley once and used the grass hay to feed their livestock over the winter. Meanwhile, the wapiti had found abundant food in their summer habitat. When the snow came earlier than usual, they migrated to the lower, exposed mountain ridges and found less forage than usual. They descended to the slopes of the valley, which had also been overgrazed by the cattle, and finally down to the valley floor, where there was nothing to eat at all. In despair, they assaulted the fenced-in hay supply of the farmers, who grabbed their rifles and shot the creatures. In

16. During the last seven hundred years, fifty-nine species of bird have died out in New Zealand and thirty-six in the Mascarene Islands [Indian Ocean]. In the foreseeable future, feral dogs and cats will have exterminated the kiwi in New Zealand and the kaguy in New Caledonia. The rich bird preserve on Lord Howard Island, 456 kilometers east of Australia, has fallen victim to rats.

the winter of 1910–11, 2,500 animals died; in 1912, there were still 20,000 head. Then, however, came the winter of 1919–20, which brought frightful new losses, and by the following spring, only 10,000 were still alive. The herd remained at this size for a while, but found itself in great danger again in 1924–25. American hunting magazines were full of pictures of the starving animals and of the vast numbers of carcasses lying around in characteristic positions. Now, people are trying to buy up the farms in order to maintain the herd's habitat. May the Isaac Walton League be successful in its efforts!

There is a European analog to this American animal drama, which Braun (25) has recently brought to our attention once again. In Sweden, there are about one hundred and seventy thousand reindeer, some of which live year round in the forest. Most, however, migrate in huge herds from the Swedish to the Norwegian side of the mountains; they are following their natural instincts, which cannot be restrained even though the Lapps and their families travel with them. To quote Braun: "The reindeer do not understand at all that a line marked out with stones or rivers is a boundary that can only be crossed in accordance with certain rules." The animals depend on snow-free places for their calving, which takes place in early spring on the eastern slope of the mountains. Afterward, they head boisterously for what the Lapps call "the kingdom the the sea"—the Norwegian coast—where, at that time of year, the vegetation is already green and the ocean wind protects the animals from the flying insects of the warmer interior. During the summer the reindeer feed on grass and herbaceous plants and, after having laboriously crossed the mountains, naturally do not pass up the meadows and pasturelands of the inhabitants. In winter, however, they eat reindeer moss, a lichen that is soft at that time of year, and the animals can scratch through deep snow to get it if the crust is not too hard. Reindeer moss grows in the interior, and so, in August and September, the herd begins to migrate back across the mountains to Sweden.

The interior of Lapland is closed to new settlers by administrative order and reserved for Lapps. Outside this boundary and more and more within it, new farms and groups of farms are being settled. The settlers, often Finns, raise reindeer themselves, do some planting, but primarily raise livestock (cattle and sheep). They need meadows because they need hay for winter feed. Hay production takes place on moors and along riverbanks, which are widely dispersed and far from the farms. It is stored on racks or in sheds until winter, when the farmers pick it up with sleds. This practice brings them into severe competition with the wild herds. In 1925, the summer in Lapland was dry and hot, and little grew. In August, it began to rain, and the scanty hay harvest could barely dry out. At the end of August and the beginning of September, the first snow fell above Lake Kilpis, in Finland, at eight hundred meters. On September 9, everything above eight hundred meters was heavily covered with snow, and a little later it snowed on the valley floor. Meanwhile, it alternately thawed and froze, not typical for a normal winter. The snow became crusty, the most difficult situation a herd can face, because then the animals cannot get to the lichens. Braun went on to say, "So then thousands of thundering, half-wild animals break into the hay storage and steal the scanty winter hay."[17]

Winter cold can cause lack of food just as drought does. It also causes great mass migrations in the animal world. The example of the wapiti reminds us of the vast migrations of the American bison, whose original territory extended from Great Slave Lake in Canada to southern New Mexico, and from Pennsylvania and eastern Georgia to Arizona and northern Nevada. The bison had a vast range, and was characteristic of the treeless plains of the West as well as of the thick forests east of the Mississippi. Its regular movements from one grazing area to the next were adapted to the changing seasons. The onset of winter in Montana and Canada and the beginning of the extension of the snowcover set the migration in motion. In the spring, the return migration from south to north took place. By bringing masses of bison together from a wide area, these regular, life-sustaining migrations also brought about the surprisingly swift destruction of an estimated thirty million or more head. First the animals were forced west, back across the Mississippi, and then their migration trails were intersected by the first transcontinental railway, which impeded or cut off their punctual march to the south. Today most of their deeply trodden trails have disappeared, but even though many highways and railway lines follow their lead, we seldom give much thought to the shaggy pathfinders that established the routes.[18]

15. DEATH DUE TO COLD

The catastrophe at Smithers Lake, which forms the paradigm around which these paleontological deliberations are grouped, is primarily a syn-

17. In 1836, in North America, below the sixty-fifth north latitude, King saw at least twenty thousand caribou pass by on one day. In North America, these unusually regular migrations can cover a distance of fourteen hundred kilometers, following the same trail year after year. Every summer, the reindeer of Kolyma [USSR] leave their forest habitat and head toward the far northern tundra on the shores of the Arctic Ocean, although some go to the mountains where the forest protects them from the swarms of mosquitoes. In the autumn, the reindeer return to the forests in great herds, swimming across rivers at exactly the same places year after year. Hunters have taken advantage of this behavior since the beginning of time; Pfizenmeyer describes how they would fall on the herd when it reached the middle of the river.

18. Herds of bison move parallel to each other: those from the north never cross the Republican River to the south. Devastation of vegetation by grasshoppers can also cause a herd to migrate. Even though bison migrate in the winter, they do not leave the area where winters are snowy, and although they try to evade snowstorms, many herds perish in them.

chronous mass death caused by an abrupt drop in temperature. Death from cold brings about many a vast carcass assemblage. The deciding factor is not the actual low temperature; it is the sudden drop that kills. Mass death due to cold is not a regular, predictable, seasonal phenomenon, but one of those events that does not fit into the limited 365-day cycle within which we humans live out our lives and make our weather observations. From the human perspective, it is "catastrophic," as are all the more severe manifestations of nature's might, which, although inevitable, do not happen regularly, do not always strike the same places, and are not of the same extent; the intervals when we are not afflicted suffice to dim our memories of the event. Such cold "catastrophes," floods, fatal droughts, storm tides, volcanic and seismic paroxysms, hurricanes, epidemics, and similar occurrences cannot take place at regular intervals because these occasional, extremely severe, and conspicuous natural events are usually the cumulative result of a complex system of interlocking factors that, taken individually, would produce effects of widely varying intensities.

In 1886 and 1894, when temperatures dropped to $-7.5°$ and $-7.2°$ C, respectively, great numbers of fish in Tampa Bay [Florida] froze to death. How unstable is the distribution of heat and cold on the earth, at sea as well as on land, and how much it can vary from the expected statistical norm we learn from a study of the climatic peculiarities of particular parts of the earth that are exposed sometimes to continental and sometimes to oceanic influences. The distribution of land and water, the levels of water, the presence or absence of watersheds, the distribution of humidity (of enormous importance to life)—all these are factors I only touch on here. The great warming influence of the Gulf Stream on the possibility for intensive settlement of Europe remains one of the most extreme examples of the remarkable and complex processes of the storage and transport of heat.

Certainly, species of animals and plants are capable of surviving catastrophes, even though many die. The survivors repopulate the depopulated area with surprising speed; on the other hand, these devastating events must not happen too often. The pauses between have to be at least long enough for the new offspring to reproduce. If this does not happen, the consequence is serious displacement of geographic range and in conspicuous instances, extinction. Displacements are regional when the creatures can evade the situation, but complete when escape is not possible. In such instances, Walther (171) uses the word *anastrophe,* which emphasizes not so much extinction as fundamental biological change in the local faunal composition. We have here, without a doubt, a principle of great significance in the establishment of trends in the alteration of animal and plant ranges. The idiosyncrasies of the changes from warm to cold and back again make these alterations erratic, unstable, and unpredictable.

Cold sets an absolute limit to life. Therefore we first spend a little time with the concept "death due to cold." Understanding this pertinent, significant phenomenon is made much easier by the fact that the relationship of migration and remigration and of dying out and complex repopulation to the multiple advances and retreats of glaciation in America and Europe has long been familiar to scholars. The animal geographer, the botanist, the student of prehistory, and the anthropologist—all must pay attention to the significance of these phenomena. If we take the literature on bird migration, we immediately have a vivid example before us.[19]

Isolated events follow each other imperceptibly until definite alterations result. We concentrate, however, on the isolated events. The geological record of these events, or better yet, the record of their frequent repetition, will finally allow us to understand more fully many events in the history of the earth. Unusually heavy snowfall and extremely severe winters may be one-time causes of destruction and mass death;[20] they may also, however, be one of a series of events that continually diminish the ranges of certain creatures. The fundamental difference between the flora and fauna of Europe and that of North America is based on these events. In both instances, advancing glaciation pushed the fauna to the south. But in Europe, the mountainous barrier of the Alps, with its own centers of glaciation, lay to the south, and only partial escape to the east and west was possible. That is the reason for the extreme complexity of eastern and western migrant flora. It is also the reason that certain individual groups of plant species have migrated more than once, at different times; that plants of the steppe, forest, and other ecologically important environments are so bizarrely and complexly distributed; and finally, that the great, homogeneous migrations are missing from the picture. Furthermore, the possibility for animals that had become adapted to the cold to migrate back north was available only on a limited basis, so most of them perished too.

The flora and fauna of America give the impression of being more an-

---

19. Bird migration in Europe even today follows three main flyways: the west coast, the Italian-Spanish, and the Adriatic-Tunisian. Less noticeable are the migrations of northern bats to winter in warmer places. Hesse reports: "*Vesperugo nilsoni [Eptesicus nilssoni]* migrates from northern Iceland south to winter in Silesia, Mähren, upper Franconia, and even the Alps. In the autumn, *V. daycneme [Myotis dasycneme]* leaves the north German lowlands to seek shelter for its winter sleep in caves in the mountains of central Germany. In America, bats, especially those that live in trees, migrate as far as the Bermuda Islands."

20. Military campaigns have often exposed great masses of men and animals to death in snowstorms. We think of the war between Sweden and Norway and of Napoleon's retreat from Russia. This was also the reason why the Russian invasion of Turania was at first unsuccessful. General Perowski's expedition against Khiwa, undertaken in the winter of 1839–40 with twenty thousand men and ten thousand camels, encountered a blizzard on the steppe between the Caspian and Aral seas, and most of the troops and animals died.

cient than those of Europe. A walk through the subtropical virgin forest of Louisiana reminds one exactly of the Eocene in what is now central Germany. *Smilax* tendrils impede entry into the woods, and their thorns tear at the skin. Palmetto forms the undergrowth. Pines inhabit the leached, washed-out prairie soil that is not exposed to flooding, evergreen oaks border the dry land, and *Taxodium* swamps hug the high waterline. Magnolia, catalpa, sweet gum, and sassafras are suggestive of Tertiary forests in Europe. Although the lack of mountain ranges stretching east and west often allowed unobstructed passage of cold waves, it also made escape to the south much more feasible. The richness in species of American trees and the retention of many Tertiary forms is, however, also due to the more peaceful geologic development of North America as compared to that of Europe. The almost homogeneous band of forest that circled the North Pole in the Tertiary was shattered by the Pleistocene glaciation. It is best preserved in America, and in addition, reverse migration from south to north is clearly seen. Trees also grow much faster on the Gulf Coast than in Europe. In summer, the Carolina parakeet goes as far north as Pennsylvania, and the hummingbird advances beyond St. Louis. *Amia* and *Lepisosteus,* the alligator, snakes, and many species of turtle are reminiscent of the European Tertiary. Complex animal and bird migrations are relics of extensive glaciation.

Too much attention is paid to degeneration and too little to regeneration, the process that over a long period of time maintains a balance with the effects of anastrophe. Intermittent, sudden expansion of populations and reconquest of lost territory is seen just as often as are the effects of castastrophe.

At the beginning of the glacial epoch, the hippopotamus was still living in northern Italy, on Crete, and even on what is now German soil. The huge turtles and pygmy elephants on Malta had not yet been exterminated; lions, wild bulls, and antelopes lived in Greece. Frech (53) once compared in detail the literature on the reasons for the dying-out of animals before and during the Ice Age, in order to compile in one work the wealth of information available on this subject. The general results of this research are presented here, although they are already a bit out of date:

A. General Results
   1. Far-reaching climatic and geographic changes brought about extensive destruction of the biological world in the geological past, thereby creating space for new forms.
   2. Immediately after the disappearance of the old plant and animal world, a new one appeared, adapted to the changing conditions and almost always of a higher order of development.
   3. More than anything else—provable only three times (Dyas [Permian], Upper Cretaceous, Quaternary)—the refashioning of the

animal world coincided with extensive glaciation or periods of cooling.

4. External causes for the dying-out are present in sufficient diversity; internal causes such as gigantism or overspecialization come temporarily into play.

B. Reasons for the Dying-out of Quaternary Mammals

During the Quaternary cold period in the temperate and polar zones, the large animals (*Ursus spelaeus, Elasmotherium,* and the giant deer) died because they had become over specialized and, therefore, incapable of adaptation.

1. When the warmth disappeared at the beginning of the Quaternary, so did the tropical and warm-temperate forms: *Hippopotamus major* in Europe; *Rhinoceros mercki* Jaeg., the immediate descendant of an Italian Tertiary species (*Rhinoceros etruscus*); and *Elephas antiquus,* immediate descendant of the late Tertiary species *Elephas meridionalis;* and, finally, the giant beaver, *Trogontherium,* and *Elasmotherium,* the largest and most peculiar representative of the rhinoceros family, from the area of the Volga [USSR].

2. As soon as higher temperatures prevailed after the melting of the glaciers in Europe, the mostly huge, Arctic mammals such as the mammoth, the rhinoceros (*Rhinoceros antiquus*), the giant elk, and the Old World musk ox disappeared. An escape of special significance was that of the giant elk to the sparsely wooded island of Ireland, where this mighty antler bearer later became extinct. In Europe, the appearance and disappearance of the large predators (cave bear, cave hyena, lion) depended on the migrations of their prey.

3. The preservation of some animal forms depended on the possibility of migration back to Arctic areas (tundra reindeer, musk ox). Mammoths and rhinoceroses were cut off from retreat to Siberia by intermittent flooding in eastern Russia; in a similar fashion, the ongoing formation of the Bering Sea cut off the retreat of the American mammoth herds.

The preservation of other forms depended on the possibility of migration back to high mountains (chamois, ibex, snowshoe hare, ptarmigan), as well as on adaptation: the European bison and the forest reindeer of Scandinavia and North America (woodland caribou) are descended from Arctic tundra forms and became forest animals after the Ice Age.

4. It is striking that in North America, there is not much mingling of terrestrial faunal elements of different ages; furthermore, few new species arose after the climate change (*Bison bonasus*). Both facts testify to the biological effects of the uniform climate of the Ice Age. The eventual consequence of a repetition of the "ice ages" in

North America—that is, of a lengthier glacial period characterized by significant climatic change—would have been a greater resident and migratory fauna and, more important, the origination of numerous new species during the presumed interglacial periods.

Low temperatures hinder putrefaction; consequently, carcasses of animals that died from cold may be preserved for a long time, as we know from Siberian mammoth and rhinoceros remains. Just as interesting is an instance of arthropods—the famous "Grave of Prehistoric Grasshoppers," one of the wonders of the Yellowstone Park. Millions of dead grasshoppers are enclosed here in the ice of "Grasshopper Glacier." S. Eckmann told Wiman (176) about an end moraine built of dead lemmings. Both instances were the results of migratory processions that succumbed to cold. According to Johannes Walther, on the New Siberian Islands [USSR] there are frozen masses of sandy snow that contain animal and vegetable remains and intercalations of ice twenty meters thick.

An odd form of death due to cold happens when water freezes solid. In 1873, near Spitzbergen, Nordenskjöld observed a considerable number of dead auks trapped in fresh ice. Such carcasses probably thaw out in summer and drift. In the same area, Wiman (176) time and again saw floating skeletons of fulmar[21] and auks. The carcasses do not sink for a long time because even after the soft parts have decomposed, releasing air and gases, and the skeleton is about all that remains of the body, they continue to float as long as some air remains trapped in the feather mantle, which fills with water only very slowly. According to Wiman (176, 179), swans and other swimming birds became icebound in a lake near Upsala one spring. At that time of year, they can usually free themselves before noon, but if the same thing happened in autumn, they would die. Swans dying in water that has frozen have also been reported by other observers. After a strenuous day's migratory flight, the exhausted birds alight and remain motionless for a long time. The very cold water freezes and holds them fast. People have also found mammals frozen in ice. The Jesuits Hus and Gabet once found about fifty icebound yaks in the Yangtze River [People's Republic of China] frozen in a swimming position. Death from cold also plays an important role in the plant world, and the temperature can fall abruptly even in the warm state of Florida. In 1886, when the fish in Tampa Bay froze to death during a severe norther, most of the mangroves on the west coast of Florida were also killed.

The complete lack of ancient fish in contemporary deep seas has also been related to the annihilation of Tertiary deep-sea faunas during the Ice Age (see Abel 4). When Hoernes (71) made his study, he was under the immediate influence of Boulanger's reports on the mass death from cold of *Lopholatilus,* said to have happened when a series of unusually severe

21. *Fulmarus glacialis* = storm petrel.

storms diverted the warm water of the Gulf Stream. This interesting and dramatic event has often been acknowledged but its aftermath hardly at all. Because it is of special importance for the paleontologist, we look at the fate of this tasty fish in greater detail.

Even very extensive natural catastrophes—we should really call such events anastrophes, as Walther does—are seldom capable of doing away with a species completely. The definitive example is the disappearance and reappearance of the tilefish, *Lopholatilus chamaeleonticeps,* one of the Pseudochromiidae belonging to the Acanthopterygii. Before 1879, the fish was completely unknown. In that year, however, New England fishing boats out for shellfish off the coast of Nantucket caught five thousand pounds of this creature, which they had never seen before. Although the taste of this newly discovered foodfish was not bad, people did not want to eat much of it at first, no matter whether fresh, salted, or smoked, but interest in it grew quickly. In 1882, however, ships entered the harbors of Philadelphia, New York, and Boston with the report that they had traveled through masses of dead tilefish miles wide. A spokesman on one steamer said it had traveled 150 miles through such a mass. Putting all the reports together, it seemed that an area of 5,000 to 7,500 square miles was thickly covered with dead or dying fish, exclusively of this species. The estimate of the number of individuals varied greatly between 1 billion and 2.5 billion. Finally, the enormous schools of tilefish were nowhere to be found. People assumed they had died out completely.

At first, the cause of death was unclear; the fish had shown no sign of sickness or parasitic infection. Obviously, the disappearance had something to do with a sudden displacement of the ranges of the currents that regulate living conditions in the ocean off the eastern coast of America: the ice-cold ones from Hudson Bay and the warm ones from the Gulf of Mexico. The tilefish lived at the place where the Gulf Stream meets the Arctic current and the colder deep sea; in 1879 and 1880, the waters there were warmer than usual. The fish had the advantages of both the abundant food from the Arctic currents and the warmth of the Gulf Stream, a combination at once favorable to the development of the species and dangerously unstable. A number of other animals from the lower latitudes were also found in this zone—a sort of peninsula of tropical fauna in the middle of the cold ocean. Shortly before the die-off, northern currents had driven Arctic ice into this habitat, robbing the tilefish of their life-sustaining warmth. People were concerned that the catch of this unusually valuable foodfish, so recently discovered, had come suddenly to an end, and they had to go back to catching kinds of fish that were still available. However, some ten years later, in 1892, the *Crampus* caught eight individuals. Increasingly larger catches in subsequent years showed the fish continuously winning back its lost territory. With its reappearance, advertisements encouraging its consumption were widespread, and for a while it was one of the most prized fish; it is still in demand today. We do not know how it escaped complete extinction. We must assume that some individuals probably escaped to warmer water; in any event, the population increased until it reached its former size.

Similar serious displacements of fauna have occurred on the west coast of South America. There, of course, the reverse happened: the cold-water fish, carried further north than usual by cold ocean currents, were pushed back, and the warm-water fauna was favored.[22]

16. DEATH ON ICE

There remains to be mentioned a curious group of accidents that result in carcasses being buried at the bottoms of great bodies of water far from shore. In the lake region of North America, it has often been observed that

22. In 1925, the ocean current called el Niño, which around Christmastime moves between the westerly trending, cold Humboldt Current and the west coast of South America and usually does not extend beyond the northern border of Peru, assumed unusual proportions, overtaking almost the whole course of the Humboldt Current. The accompanying rains either made the desert west coast of Peru fertile or destroyed it. Floods carried much rafted material to the sea, where it collected in bays and harbors; it contained not only vegetal debris but many dead birds and reptiles. Much of the cold-water fauna was killed. The sessile benthos was especially vulnerable, and much of it died. Dolphins and many warm-water fish moved far to the south, but the usual schools of fishes were not present, hurting not only the fishermen but also the abundant bird life of the guano islands. As Adolf Wesemüller describes, the lightship and the rocky coast near Callao were covered with birds that were wasting away because their usual food, the silvery sardine, was missing. The birds suffered so from the unusual water temperature that they were soon too weak to clamber up the rocky shore. Although there were countless victims, there were naturally also many that saved themselves by going farther south. The die-off of fish, the disappearance of the mighty schools of fish, the formation of sulfuric acid-rich sediments in the bays, and increased sedimentation at the mouths of valleys all went hand in hand. Because such events happen over and over, these enormous, extremely populous biotic communities are always in an unstable, endangered situation. With varying intensity and at roughly seven-year intervals, the same thing happened in 1878, 1884, 1891, and 1908. The geological significance of such episodic events for the formation of sediments containing abundant organisms is obvious.

Within a few decades, the fauna in lakes can change in ways that the paleontologist might think possible only over a long period of time. A.G. Huntsman ("The Quill Lakes of Saskatchewan and Their Fishery Possibilities," *Contr. to Can. Biol.,* 1922, 127–41) indicates that within historic time three different fish faunas have alternated in the Quill Lakes. The older Indians tell of a time when there were many fish. Not long ago they were gone altogether, but today, three men with a one thousand–yard net can catch almost a ton of fish a day. These lakes, which cover 230 square miles, have absolutely no outlet and are today half as salty as seawater. Water level and salinity change from year to year according to climatic conditions. On the shores of two of the lakes, there are extensive deposits of gastropods, which no longer live in the lakes but come, rather, from the area of the inlets. Sediments saturated with gastropods, in which zones of abundant fish remains alternate with zones devoid of them, formed in the lake.

game animals pursued out onto the ice by wolves may break a leg and finally die. I have a picture of a young deer that made a swift escape from wolves but lay on the ice of Lake Huron with a broken leg. Similar accidents also happen if animals are pursued into water, mud, or a river eddy. One also supposes that hunting wolves provided the victims for the animal trap at Rancho la Brea, near Los Angeles. Högbom wrote that not too long ago, a block of snow with four hundred reindeer standing on it slid into a lake in Jemtland. Wiman (176) reported three separate incidents in 1909 and 1910 in which 93, 80, and 94 reindeer died after breaking through the ice on a frozen lake in Norrland [Sweden].

3 Laws governing positions of recent and fossil vertebrate carcasses

1. BURIAL OF LAND VERTEBRATES IN MARINE STRATA

When fossil walrus teeth are found at Red Crag in Suffolk [Great Britain], as Lankester describes (98), it is not particularly surprising. It is really more peculiar that marine benches near the coast not infrequently contain remains of land vertebrates. For example, the offshore formations of New Zealand have occasionally yielded remains of the numerous species of moa (Marshall 104). Near Mukomaru Beach there are Pliocene cliffs where remains of moa bones (*Dinornis robustus*) were found together with a purely marine mollusk fauna, 8 percent of which were from still-living species.

In 1912, in the old sand grotto near Majunga in northwest Madagascar, Kaudern (85) discovered an accumulation of small land vertebrate bones among which, aside from those of species still living today, were remains of two extinct animals: a lemur and a species of *Microgale*. We have already mentioned the accumulation of pygmy elephant bones in beach caves on Malta. Tertiary islands, especially those in a Septarienton [clay of Middle Oligocene (Rupelian) age containing abundant septaria], are sometimes discernible only as marine beds containing bird bones. The isolated find of ancient flamingos (*Scaniornis lundgreni*) in the Saltholm Limestone at Limnahm near Malmö [Sweden], reported by Dames (35), is interesting. Studer (160) compared the burial of mammal remains in the marine Molasse deposits at Brüttelen [Switzerland] with what he observed on the northwest coast of New Guinea, where Salawati Island is separated from the mainland by the Straits of Galaevos, a channel that gradually increases in width from two to ten nautical miles. Along the side channels, he observed a dense mixture of remains of land vertebrates and marine fauna, exactly as in the Molasse at Brüttelen, where mammal bones and remains of land tortoises appear with shark teeth, *Cardium, Thracia, Tapes,* and so on. We often find remains of land vertebrates in the offshore clay of marine deposits; examples are the protorosaurs of the Kupferschiefer or the Eocene lophiodont remains on the basal phosphorite deposits of the Lower Oligocene of the Helmstedt trough (Schroeder 147). The occurrence of Middle Eocene mammals in the lignite of the Geiseltal [East Germany] makes the appearance of abraded teeth in the overlying transgression formations understandable without further explanation.

Klähn (89, 90, 91) thinks it is nearly impossible for the carcasses of land animals to have been buried immediately in marine sediments of the same age. Basing his thinking on a number of examples, he believes that land-mammal remains found in marine beds ordinarily come from older lignite or fluviatile sediments, where they were fossilized before being washed out by wave action. No doubt this process plays a role where coastlines are sinking, which is why we find remains of Pleistocene land fauna on top of the Dogger [Great Britain].

However, based on my studies on the Gulf Coast, I agree with Moos (116) that burial directly in brackish and marine beds can be both an isolated occurrence and the result of anastrophe. Land mammals by no means avoid the shore, not even the arid beach; they go so far as to prowl the long, narrow sandbars between the lagoons and the open sea and, although it may be somewhat dangerous, sometimes find their bread and butter there in the hollows of the dunes, where they can also dig down to fresh water. There are even herbivores that have become adapted to salt vegetation. I have observed plenty of bones everywhere on the arid beach. There, especially at night, a rich animal life unfolds and leaves its mauled prey behind to decompose quickly during the day. Along the Gulf of Mexico, we find the tracks of the bay lynx (*Lynx rufus*), an unusually muscular, agressive, nocturnal animal. It sometimes attacks the ring-tailed cat—the cacomixtle of the Aztecs (*Bassariscus astutus*). It too is mostly nocturnal and feeds on many different things. On the beach near Kingsville, I often found remains of members of this species that had been attacked. The raccoon (*Procyon lotor*) also leaves the wooded floodplains to make nightly forays to the seashore. Carcasses of opossum (*Didelphis virginiana*) are also fairly common. Added to these are carcases of fish, birds, porpoises, turtles, snakes, cattle, and horses; one can see many bones protruding from the upper layers of low beach scarps. The typical animal of the Louisiana salt marsh, the muskrat (*Fiber zibethicus* [*Ondatra zibethicus*]) is also found on the beach. (This aquatic creature, whose range extends from the Atlantic to the Pacific, is especially numerous in the deltas of the Mississippi and the Colorado. In 1914, ten million pelts were marketed and the meat sold as "marsh rabbit.")

Moos (116) describes three types of burial of land mammals in marine strata: (*a*) mammal remains were immediately washed into marine depo-

sits of the same age; (*b*) mammal remains were washed out of older limnetic or fluviatile deposits by the sea; and (*c*) mammal remains have their
primary burial in river deposits, and the marine fossils occurring with
them were washed out of older marine sediments.

Based on this assessment, Moos classifies southern German localities
[West Germany] as follows:

1. Of the land-mammal remains of the Swabian Burdigalian marine
   Molasse, those from Baltringen in upper Swabia, as well as those of
   the Swabian Jura zone east of Ulm, are probably in their primary
   depositional position. Those from the mountain road at Erminger
   were possibly washed out of the subjacent *omphalosagda* beds.
2. The middle Miocene mammal remains in the Grimmelfinger
   Graupensand, the Reyath gravels, and the coarse sands of Riedern
   and Benken lie in their primary burial in Vindobonian river gravel.
   The marine fossils found with them were washed out of older
   marine sediments.
3. Near the confluence of the Danube and the Iller are two mammal
   localities whose difference in age is slight; a lower one lies in the
   brackish *paludina* sands at the base of the Kirchberger beds near
   Unterkirchberg and an upper one in the limnetic *sylvana* beds near
   Oberkirchberg and Günzburg. In both localities the burial is
   primary. At the Geelenhof locality, it is possible to find a mixture of
   both horizons.

In contrast, Klähn (90) looks at the remains of a lower Miocene mammalian fauna in the localities at Pfullendorf, Junghof, and Hause in the
Andelsbach valley [West Germany]. These remains were processed by the
sea or redistributed by erosive rivers and deposited in the offshore clay of
the marine Molasse. In the beds underlying the Danube delta, beneath
shell banks, remains of land mammals that had not been displaced were
found—remains of mammoth and rhinoceros, in fact. Delta sediments
are often rich in remains of mammals, crocodiles, turtles, land snails, and
hard parts of marine creatures, as Andrée (9) has explained in greater
detail. Deltaic strata with vertebrate remains can also be part of large
freshwater basins.

A few pictures taken at Matagorda Bay illustrate how the burial of vertebrate remains in marine beds proceeds today. We can observe the same
phenomenon on the open gulf, especially where whole herds of cattle
have foundered in quicksand. Plate 16, figure B shows a photograph,
taken September 2, 1925, of a low, partially destroyed beach scarp on the
salt marsh belonging to the Baer Ranch, at the eastern end of Matagorda
Bay. The cow was a victim of the norther that hit at the end of 1924, and
the skin has since dried out like parchment and sticks to the bones. Within
this short time, the surf has partially undercut the place where the cow lies,
causing the front end of the carcass and one foreleg to hang over the edge

of the scarp. The tip of the right horn is pointed upward; the skull, however, has already been shattered by waves. Plate 10, figure A, a photograph I took on October 10, 1925, shows a cattle skull already almost completely buried in a beach sparsely littered with oyster shells and sargasso. It has been buried with the horns pointing downward. The occipital region and the chewing surfaces of the teeth of the upper jaw are still visible, as is the right mastoid process. The position of a skull I photographed on September 3, 1925, at eastern Matagorda Bay (pl. 9, fig. B), is similar. Only the occipital region and the rear end of the left row of teeth can still be seen. The skull is embedded at the strandline, where coarser and finer materials meet, and driftwood is lying in its flowshadow. Plate 9, figure D shows a view of the beach at the northern shore of Matagorda Bay, near its eastern end. We see remains of dead cattle, but we also notice that the shoreline is eroding. Waves gnaw away at the land, washing out the buried skeletons, which are then carried off and deposited again in deeper water. The shore goes on producing bones, and erosion goes on washing them out; in this way, a vast, bone-rich plain is formed. This process usually disarticulates the skeleton, and the lighter elements are carried farther away than the heavy ones. Plate 9, figure C shows an isolated vertebra lying in the shoredrift.

Abel (2) has often reported so descriptively about the destruction of carcasses by waves and river currents that we can do no better than to quote him:

> Individual vertebra become detached, those of the tail first, then the extremities and skull. Finally, under favorable conditions, the thorax, throughly jumbled, is buried in the bank. Under unfavorable conditions, the skeleton is torn apart and tumbled along by water until it has lost its original form. The extensively transported and abraded vertebrae and leg bone fragments that we find so often in the Leithakalk of the Vienna Basin are the product of such a process. The destructive effects of the surf are such that only the hardest part, the enameled surface of the crown of the tooth, resists for long and is, therefore, most often found.

Further:

> When the carcass of a land or sea mammal is washed up onto the beach, most of it is destroyed by the surf, and what is left is finished off by the scavengers of the sea. This explains why complete skeletons are so seldom found in Tertiary marine deposits; what we find most often in these formations are single skeletal elements such as jaws and other parts of the skull, broken teeth, isolated vertebra, pieces of broken ribs, abraded leg bones, and so on.

And finally:

A large carcass rolls along in the swell, closer and closer to the shore. It is a dead manatee, whose abdomen, blown up like a balloon by the gases of decomposition, rises now and then above the surface. Then a breaker shoves the carcass onto the sandy beach. It is already in an advanced state of decay: the head hangs loosely on the trunk, and the lower jaw has long since fallen to the bottom of the sea. Now it lies rotting on the beach where, with each little nudge from a wave, it slowly bores deeper into the soft, damp sand. After a few days, a high tide will cover it with a layer of sand, thus protecting it from further destruction.

Destruction of the skeleton by surf is typical for sirenians from the Mokkattam stage of the Eocene.

Remains of mammals from different ages may easily be found mixed together on the coast, exactly as are the hard parts of invertebrates of different ages. In riverbeds and on sandbanks, we often find remains of Eocene mammals next to those of contemporary fauna (pl. 10, figs. B, C, D). In the ichthyosaurs of the Posidonia Shale, we often see submarine disintegration caused by complete maceration. Von Ammon (15) has already shown that in the Solnhofen lithographic limestone, the fish near Nusplingen are more completely disarticulated than those near Solnhofen; furthermore, in contrast to Solnhofen, Eichstätt, Pfalzpaint, and most of the other Altmühl lithographic limestone localities [all West Germany], at Kelheim the fish and reptile carcasses are severely disarticulated. He maintains that at Kelheim and Nusplingen, the covering with mud took place more slowly than in the other places. Rising water levels in that area would also account for these findings. Ammon (15) emphatically denies the interpretation that "at this particular spot the carcasses washed up and deposited on the mud flats were in a more advanced state of decomposition, while in other places only living or better preserved (less severely decayed) specimens ended up in the muddy coastal lagoons."

An interesting example of complete disarticulation is offered by a skeleton of *Homoeosaurus maximiliani* (HDE) (see fig. 3) described by Ammon (11). The carcass has come completely apart, and the separate elements have been carried off, perhaps by scavengers; some parts are missing. We know of quite a few specimens of the same saurian in good, completely preserved condition. Water currents and gravity moved the bones around somewhat, and then these complete skeletons were buried soon after the river dried up. The carcasses did not dry out, but rather decayed in shallow water.

Johannes Walther (169) observed that the *Aspidorhynchus* in the Solnhofen lithographic limestone at Kehlheim most often occurs disarticulated; often only the head with connected intestine is preserved. The fish is common at Eichstätt but seldom found at Solnhofen, Langenaltheim, and Daiting. Heineke (64) shows in his figure 8, page 29, and in plate 2, figure 8, all the parts of the broken skull of *Aspidorhynchus acutirostris* with the long parasphenoid in the middle. According to Walther, out of six specimens of *Spathobatis mirabilis,* only one had been attacked by decay. He thinks the animals must have died a peaceful death. Several fragments of *Ischyodus avita* Meyer were found at Kehlheim; they had partially decayed and had been torn apart by the surf. *Eugnathus macrodon* always lies in a curved position. *Caturus furcatus* is well preserved at Kelheim and Eichstätt but mostly disarticulated at Solnhofen. Almost all the specimens have food fish inside them. *Strobilodus giganteus* Wagner is seldom found at Eichstätt, Solnhofen, and Nurtingen and more frequently at Kehlheim, but in a worse state of preservation. *Leptolepis sprattiformis* [*Leptolepides sprattiformis*] is especially numerous in the underlying beds at Langenaltheim, where it occurs in several horizons. At Eichstätt it occurs once in a while. *Thrissops* usually occurs individually and almost always has a small *Leptolepis* in its gullet. Of the pterodactyls, usually the *Rhamphorhynchus* are completely preserved. A skeleton of *Pterodactylus kochi* Wagner [*Germanodactylus kochi* (Wagner)], of *Pterodactylus propinquus,* and of *Pterodactylus ramphatilus* Wagner [*Germanodactylus ramphatilus* (Wagner)] were found disarticulated; of the latter, according to Walther, all that was found was a pile of bones from the collapsed spine.

In some deposits, the skeletons are always found disarticulated. The batrachian remains in the Skiritz lignite, near Brux [Czechoslovkia], described by Laube (99), are an example. The bones have become completely separated from their ligaments and have been distributed in the shaley stone in such a way that they are usually visible only individually. When the skeleton happens to fall apart, and the bones dissolve (a process that takes time), only especially resistant parts are preserved. Thus, the sturdy tympanic bullae of whales, the enamel surfaces of sharks' teeth, and the otoliths of bony fish are almost the only evidence we have enabling us to identify the fish fauna in the Neogene of the Vienna Basin. Schubert (145) used them to identify the coastal fauna at Steinerbrunn, Gainfahrn, Enzesfeld, Perchtolksdorf, Grinzing, Nussdorf, Seelowitz, Vöslau, Krenberg, and Neudorf an der March; he identified perch, brace, gudgeon, plaice, sea robin, and drumfish along with isolated deep-sea forms. The marl at Baden, Boratsch, Neudorf, Lapugy, Möllersdorf, Traiskirch, and Niederleis contains Scopelidae and Gadidae. This is, of course, a deep-sea fauna. Walbersdorf has a rich fauna of deep-sea Gadidae, especially *Macrurus.*

2. THE LAW OF THE LOWER JAW

Even in instances in which the carcass is already firmly anchored, we have seen that parts that are bulky but light in weight and do not have an extensive basal area can still be moved this way and that. The peculiar handle-shape of the lower jaw, the large hole in the skin formed by the mouth opening, the easy detachability of the symphysis, and the structures in the neck—all favor to an unusual degree the splaying of the lower jaw. Next to the anal aperture, scavengers' preferred place of attack is the mouth; by tearing off the lower jaw, they can get at the gullet and clean it out. In most cases, the lower jaws of carcasses floating in water are either nibbled off or are lost through decay. No wonder, then, that the lower jaws of fossil skeletons are often askew or missing altogether. I have only to recall the Tübingen slab, described and drawn by Fraas (51). Cuvier (34) shows in volume 2, figure 1, plate 234 a "gavial from the limestone schist of Monheim in Franconia." The lower jaw, with both rami still attached, is tipped over and lies between the skull and the spine.

Another example is the skeleton of a fossil porpoise from California, described by Lull (103). It is moderately disarticulated; the lower jaw still hangs from the ligaments but is splayed at an angle. We have mentioned the splaying and removal of the lower jaw of pterodactyls elsewhere. Ammon (12) describes a *Rhamphorhynchus longicaudus* that shows this phenomenon *in statu nascendi,* so to speak. In fossil material, we often see the displacement of the the two halves of the lower jaw due to pressure from the upper part of the skull and dissolution of the ligaments of the symphysis. *Melosaurus uralensis* from the Permian of the western Urals (Meyer 109) shows both rami so folded down to the horizontal that their lower edge is turned outward; the teeth, therefore, also turned ninety degrees, point toward the teeth of the upper jaw. The lower jaws are so splayed that the wider end of the skull lies exactly between them. Another stegocephalian, *Actinodon frossardi* Gaudry (Zittel 185) from the Permian near Autun in France, shows the left ramus in the position just described, but the right one tipped over with the teeth pointing outward; the skull lies over it.

Lower jaws that have fallen off are often sturdy and not easily destroyed. Once separated from the carcass, they can be transported quite a distance and are often found in fairly coarse sediments. What comes to mind here is the lower jaw fragment from a *Mastodonsaurus* from the middle of the main conglomerate at Altensteig, in the Württemberg Black Forest [West Germany], described and illustrated by Schmidt (142). Vertebrate rami are often found on beaches. Plate 10, figure E shows the lower jaw of a cow. It is lying with its outer side down; the concave side with the teeth is turned around toward the shore, the convex side toward the ocean. Because the jaw forms an obstacle, the volume of water flowing ashore at this point is constricted, thereby intensifying the flow, causing it to form a little channel that follows the contours of the jaw as far as the maximum curvature of the ramus, where it radiates away from the edge at right angles to the bone. The lower jaw of rays can be very durable; I recall one from *Janassa bituminosa* described by Jaekel (80).

3. THE PASSIVE POSITION OF "WATER CARCASSES"

The possibilities for preservation under water are so much greater than under subaerial conditions that remains of land vertebrates usually occur only as "water carcasses." Under water as on land, gases from the intestine and putrefactive processes gradually fill the carcasses within a few hours after death, causing them to rise again to the surface. In still waters they drift ashore, and in flowing water they are carried off and broken up, and may later collect on suitable sandbanks and meanders. This resurfacing is often prevented by predatory fish or, even more often, by crocodiles tearing the carcass open. When carcasses drift ashore one by one at the same place, nonsynchronous carcass assemblages can form.

Therefore, we often find slightly damaged water carcasses in a completely passive position similar to the natural structure of the body. Let us begin with mammal carcasses. Plate 10, figure F shows a recent water carcass, an opossum partially buried in the wet sand of a beach. The extremities are extended but slack. The forelegs are already somewhat embedded; the hind legs are still free. The muzzle, in front of which lies an oyster shell, is slightly open. The spine is so curved that the fore- and hind legs are touching. Two large bunches of sargasso lie in the flowshadow created by the carcass. The beach is thoroughly wet and not very firm under foot, a situation that causes the carcasses to disappear relatively quickly into the shifting sand.

We find this strictly lateral, passive position of the opossum carcass extraordinarily often in fossils. We need think only of the well-known skeleton of *Phenacodus primaevus* Cope in the lower Eocene of the Wasatch Beds, in Wyoming. The fossil mammals in the Molasse at Oeningen, described by H. von Meyer (107), lie in the same position. About the skeleton of *"Canis" palustris* [species probably now *familiaris*] shown in his plate 21 (see pl. 37, fig. B), he writes as follows:

> This animal, which has come down to us completely preserved, lies with the right side on the stone and the left side exposed. The legs hang down slackly beneath it, the tail is pointed somewhat upward, and its whole position gives the unmistakable impression of a dead body that lay on the ground before being covered by sediments, and not of an animal that while still alive fell into a swamp or marsh and died there.

The same author says about *Lagomys oeningensis* [*Prolagus oeningensis*], referring to his illustration in plate 3, figure 1 (see fig. 2): "The creature has been exposed from above. The legs hang slackly from the torso." According to Haupt (63), the propalaeotheres [*Propalaeotherium*] from the Eocene bituminous shale at Messel [West Germany], are in the same position. The Messel animals, unlike those from Buchsweil, did not decompose before burial. As to the cause of death, Haupt comments as follows:

In my paper on the *Propalaeotherium,* in referring to the manner of death I said that while drinking, they must have fallen victim to the countless crocodiles that, according to the remains we find, lived there at that time. I would like to modify this view, for while this manner of death may have occurred in individual instances, the majority of the deaths are mainly attributable to other causes, two in particular. The first is surely that severe storms, accompanied by cloudbursts such as occur still today in the tropics, led to a sudden flooding of the low-lying, swampy wood, bringing death to many of our ancestral horses. Then, too, predators may also have been responsible for their demise. So far, however, we know of no carnivores from the Messel deposits, but there must have been some among the fauna of that time. Frightened by attacks from that quarter, herds of propalaeotheres and lophiotheres ran blindly into the swamps around lakes, where they sank in under their own weight. They could not extricate themselves, and in fact their efforts to do so caused them to go in deeper and deeper. The sapropel swamp preserved their carcasses so well that we find, in some instances, the still clearly recognizable carbonized outlines of their bodies and hair.

We should notice, however, that the position of the animals he is speaking of does not correspond to a sinking down into the ground. They have not sunk in, but drowned. Crocodiles probably tore into their bodies, preventing them from rising to the surface again. The position can best be seen in Haupt's illustration of *Lophiotherium messelense* [*Propalaeotherium messelense*] in his plates 28 and 29. Five female and ten male propalaeotheres have been found lying in the same position. One still has its milk teeth, two are in the process of changing their dentition, and only one, according to the wear shown by the teeth, is very old; eleven, on the other hand, died in the prime of life. These examples are sufficient to prove the point.

Cuvier (34) found some good specimens of bird carcasses in the passive position in the Tertiary of the Paris Basin. In his volume 1, plate 154, figures 2a and 2b, he shows a bird skeleton whose wings and legs are partially flexed. The bill and the legs, somewhat askew, point to the same side. The wings in his plate 155, figure 1 are more spread out, and the neck and head are extended.

Fossil reptiles furnish us with many, many examples of the typical passive position of water carcasses. The first that come to mind are the Solnhofen saurians. The dorsal or ventral position is the norm; lateral positions seldom occur. The throat and tail are either extended or only slightly curved. The forelegs angle to the rear, the femurs usually bend at right angles, and the tibias converge again somewhat toward the rear. Many also have a slight curvature, immediately noticeable in the position of the ribs on both sides of the body. Sometimes the legs on one side are extended more than those on the other, but in general, the picture is the same. In lateral positions, both the fore- and hind legs point slackly toward the rear. The carcass of *Geosaurus gracilis* shown by Abel (3) lies with the legs bent toward the rear and the pelvis in a lateral position; the head, however, is upside down, so the moment of rotation is in the vertebrae of the neck. On the subject of *Homoeosaurus,* see Meyer (110), Zittel (185), and Broili (26). The bowed curvature of the long-tailed lizardlike *Pleurosaurus* (pl. 36, fig. B) from Sappenfeld, not far from Eichstätt, also corresponds to a natural, passive death position, which Broili (29a) has just compared to the death position of recent lizards of similar appearance killed in alcohol. The passive position of fossil frogs with their partially flexed legs (*Rana danubia* var. *rara),* from the Upper Miocene of Steinheim [West Germany] as described by Fraas (52) or, from other localities, by Wolterstorff, is quite typical.

The passive position of fossil snakes often consists of several slight, intersecting loops. Meyer (110a) shows a picture of *Coluber atavus* in which the twisted spine forms three short, very even curves. Nopsca (121) describes how the ribs of *Pachyophis* on the concave side direct the curvature more strongly backward than on the convex side. In his main example, the opposition of the ventral position of the torso to the lateral one of the neck is striking. Janensch (82) contains an unusually interesting description of the position of the intersecting loops of *Archaeophis proavus* (see plate 36, figure A) from the Eocene limestone of Monte Bolca [Italy]. The position of the snake is predominantly lateral, since the ventral side is clearly too narrow. At short intervals, the position of the torso changes from one side to the other. At the places of greatest curvature, the ribs, acting as a lever, pull the vertebrae around their own axis so that they turn partially to the dorsal side. There are also many elongated and needle-shaped fish in this somewhat curved, passive death position, such as *Belonostomus tenuirostris* Ag. from the Solnhofen lithographic limestone (see fig. 4), as Eastman (44a) has shown. The somewhat curved position of the teleosaur from the Posidonia Shale at Holzmaden, shown in Hauff's illustration (62), can probably be explained in this way too.

On such elongated forms, one often sees a different state of preservation at each end of the body. Another specimen of *Coluber atavus* from the lignite of the Siebengebirge [West Germany], depicted by Meyer (110a), shows the front half of the body still articulated, whereas at the rear, the elements have become completely separated. This is reminiscent of the condition of the plesiosaur *Plesiosaurus guilemi imperatoris* [*Plesiosaurus guilemiimperatoris*] described by Dames (36):

> In sharp contrast to the orderly position of the spinal column of the back is the disorder of the tail, the change occurring immediately behind the pelvis. From that point on, the vertebrae are still somewhat in contact with each other as far as the middle of the tail, but the direction of the spinal processes, the diapophyses, the hema-

pophyses, and also the centrums is very different in each vertebra, as a glance at plate 1 shos. The rear half of the tail is separated from the front half by a considerable space, in which lies a single vertebra, displaced toward the upper left. Above and below the space, far from the spine, lie three vertebrae whose separation and displacement account for the space. The forward part of the rear half of the tail is more disorderly than the forward half, while the rear part remains much more undisturbed. The slab around it is dark and greasy looking—the remains of the tail fin that will be described in more detail below. This fin may be the cause of the state of preservation just described. Because it was big enough and flat, it was possible for waves to set this part of the carcass, but not the heavy torso, in lively motion, and so, under the combined effects of wave action and decomposition of the spinal column, the caudal vertebral column collapsed into its separate parts at the point where we see the disturbance today. The tip of the tail was more protected by the fin surrounding it.

4. DISPLACEMENT OF CARCASSES

In the paleontological collection of the old Akademie in Munich, there is an interesting skeleton of *Homoeosaurus brevipes* H. von Meyer from Solnhofen. Rothpletz (135, p. 313, pl. 1, fig. 5) has described the piece in detail (see fig. 1; also see Abel 3, p. 462, fig. 388 for comparison). The completely preserved skeleton lies on a solid slab of limestone. The legs on the right side are spread out flat on the slab; those on the left side, however, go down into the sediment. Most striking is that next to the torso and head of the creature, there is an impression on the slab that very clearly depicts the outline of these parts of the body. Rothpletz (135) gave the following explanation:

> When you look at this slab, you seem to see the death struggle of this reptile before your very eyes: It suddenly came upon a dewatered area where the mud was still soft, and sank in. Using its strong tail for support, it threw its head upward in an effort to escape, but in the process the tail pressed backward deeper into the mud. It managed to free its head and torso, but they sank in again and were caught fast, as were the legs of the left side. It finally suffocated in the mud. But the depression it had created when it sank in the first time did not disappear, because the mud, no longer fluid, quickly solidified.

Two objections to this interpretation of our fossil and its death struggle can be raised based on the following considerations. First, we can ask how a living terrestrial reptile could have penetrated so far into the sea or lagoon without having sunk into the mud much earlier in its journey; and second, why it had set out on such a long migration in the first place.

In any event, dry land lay far from where we found our *Homoeo-saurus,* and, if it was not a marine reptile, it must have wandered for miles across dried-up ocean floor to reach the place. Its feet and legs lead us to the conclusion that it could have lived on land, but they do not tell us whether it had webbing between the toes, which would have enabled it also to swim. In the latter event, all difficulties would be set aside, and we could assume that this animal swam out to sea to find food and as the water receded, found itself ashore, where it sank into soft mud and suffocated.

If we dispute its ability to swim, then obviously there is only one extremely unlikely supposition left—the lagoon dried up, the mud crusted over, and the lizardlike animal hunted its prey there until it came to a spot where the ground was still wet. However, there are no indications that such a drought really occurred at that time, and therefore it cannot be said that this kind of reptile is terrestrial.

There is, however, still another possibility that Rothpletz does not mention. According to everything we know today, the animal was really terrestrial, and it is entirely possible that it had already drowned before it sank in. It is possible, therefore, that the depression following the body outline formed post-mortem. Without belaboring the reasons that speak for or against this point of view, I would like to say that I have quite often seen such postmortem displacement of carcasses, especially on the seashore.

On February 25, near the southern Texas town of Sargent, south of Bay City, in Matagorda County, I observed the carcass of a juvenile porpoise (pl. 11, fig. A). The creature was about one meter long, was in a lateral position, and exhibited a clearly visible, typical S-shaped curvature. The tail pointed upward, and the head curved toward the abdomen. Tracks in the sand show that birds had fed first on the ventral side, but so close to the middle of the carcass that they had no effect on its position. There was another feeding site on the skull above the snout. Every time the birds tried to pull out pieces of flesh at that point, the carcass, whose axis lay at the thickest part of the body—the neck—swiveled. (The pivotal point probably coincided with the center of gravity, an important factor in the anchoring of carcasses that have drifted ashore.) The head went in the direction of the pulling, and the tail in the opposite direction, and the resulting drag mark overlapped the tracks made by the gulls. Every tearing movement of the birds corresponded to a step in the drag marks made by both the tail and the tip of the snout. Visible in the picture are some twenty stages of movement in the drag mark of the tail and twenty-three in that of the snout. The swiveling describes an angle of sixty-five degrees.

Plate 11, figure B shows an odd pelican carcass I observed on May 5, 1925, on the Gulf Coast between Sargent and Mitchell's Cut. The pelican was beached after drifting awhile in calm water. Waves had opened the bill and crammed the pouch full of sand, enlarging it unnaturally. The veins stood out clearly, and when I opened the pouch, I found they had left their impressions on the mass of sand inside. Between the throat pouch and the body, a hollow had formed. Other birds had been feeding in the area of the cloaca and in the process had shifted the carcass around so that the body pointed toward the beach, and the head and neck pointed seaward; the movement swept the seaweed (sargasso) in front of the bill around too. Thus, a hollow corresponding to the creature's original position, bounded by its carcass on one side and the impression of the outline of its carcass on the other, had formed. The angle formed by the pouch and shoulder is clearly duplicated in the sand. This example is reminiscent of the *Homoeosaurus* of the Solnhofen lithographic limestone and is also interesting supporting evidence for the formation of steinkern and casts of typical body cavities.

Plate 11, figure C shows the opposite situation—a carcass that was too heavy to be moved by the scavengers in question. It is the carcass of a shark about one and a half meters long, which I saw on May 7, 1925, on the coast of the Gulf of Mexico. Gulls had succeeded in opening it on the left side by pulling out the fin behind the gills. The sand on that side was soaked with blood, and there were fresh tracks in it, whereas on the other side of of the carcass there were only older, smudged, indistinct tracks.

Finally, here is another example of carcass displacement (pl. 12, fig. A), not by the ocean but by fresh water, at the large carcass assemblage at Smithers Lake, about which we will presently speak in detail. It is the carcass of a large alligator gar in a ventral position with typical ganoid curvature. The creature had originally lain in water, the level of which receded only slowly. Consequently, vultures had begun to light, first on the back of the tough, elastic, smoothly coated, almost invulnerable carcass, where there was really no good place to stand while feeding. The flecks of mud there are the impressions of their muddy feet. They also tried standing in the mud to feed from holes torn in the scaley hide. Their desire to feed must have conflicted with their fear of unstable ground, a dangerous situation, which probably occurred often enough at the asphalt lake at Rancho la Brea. We see some tracks impressed deeply into the mud near the feeding point, indicating the degree of leverage required to pull out the flesh. Finally, the carcass had swiveled approximately thirty-five degrees, creating a smooth track and leaving the space it had originally occupied exposed.

5. FEEDING GROUNDS

Crocodiles, at least as common in earlier times as they are in many areas today, have left their characteristic stamp on many fossils. Although they are voracious predators underwater, they also tear into sinking carcasses, making it impossible for putrefactive gases to cause them to rise to the surface again. Many a hippopotamus hunter has been cheated out of his catch in this way. It has also been repeatedly observed that crocodiles are capable of dragging into the water heavy carcasses that have been lying on the bank. Nopsca (120) has recognized the importance of crocodile feeding for the formation of bone assemblages. He construed the accumulations of saurian bones in the Middle [Upper] Cretaceous limestone of Transylvania [Romania] as feeding grounds of crocodiles. There, remains of terrestrial saurians and crocodiles are found together; we see similar deposits in the Gosau Formation [Austria], and we almost always find turtles at such places. Feeding grounds also seem to occur in the northern German Weald. Whereas animals have been buried in the upper sandstone complex without having been torn apart (Koken 92), the remains at the top of the coal seam and within the seam are in much worse condition. Isolated teeth, skeletal elements, and armor plates, most without any indication that they are associated with each other, are lying all mixed together. Abel (4) says that the Miocene of the Steinheim Lake is a crocodile feeding ground where pelicans, flamingos, herons, ibises, geese, ducks, three-toed horses, pigs, rhinoceroses, mastodons, deer, and predators have been torn apart by crocodiles. The same author also mentions the large prosimian skull from the Quaternary beds of southern Madagascar, which has distinct round holes made by crocodile teeth in the cranium.

Hummel and Wenz (78) also interpreted the brown coal clay of a filled-in maar in the northern Vogelsberg in the same way. There, in bore holes, were found small teeth and jaw fragments from moles and a few other small mammals, and frog bones and turtle shells. The occurrence of crocodile teeth and coprolites, the latter containing many bone fragments, leads us to suspect that these creatures were the ones responsible for all the other bones. (According to my observations, alligator coprolites are relatively free of bones because these animals regurgitate much of the residue of their food.) The deer remains in the Late Tertiary of Ober-kirchberg [West Germany] lying above the true Kirchberg beds have been buried in such an extremely fragmented condition that one concludes that they were gnawed to pieces by the numerous crocodiles. The beds in question are part of a freshwater formation of the *silvana* zone (Moos 116).

Compared to the crocodiles' agility in water, their clumsiness in moving about on the mud is striking. They often inhabit shallow lakes that recede quickly, leaving behind a wider and wider band of drying shoreline, an interesting zone rich in animal tracks and other desiccation phenomena. It has been observed that crocodiles lay their eggs as far as three kilometers from the shoreline of a lake, anticipating that the water level will rise. As fish die on their way to breeding grounds, so do reptiles. We can compare this situation to the description in chapter 2, section 13, of the effects of drought and, in particular, of the devastating effects of a long one in the area around the mouth of the Amazon.

In the shaley coal at Messel, evidence of activity of the alligators often found preserved there is occasionally observed. At least one species of alligator, *Diplocynodon haeckeli* von Seidlitz (Seidlitz 150), has also been found in the central German Eocene lignite swamps at Sieglitz, near Kamburg an der Saale. The excavations of Eocene mammal remains, made in the Geiseltal at Walther's instigation, led to the discovery of typical alligator feeding grounds. Here are found not only massive quantities of *Diplocynodon* teeth, but also alligator droppings in abundance, which look just like the recent alligator excretions I saw repeatedly in dried-up river branches along the Gulf Coast. In one of these Eocene coprolites, gizzard stones were found. Today we know that the vertebrate fauna of the Geiseltal included nearly thirty species, but all the remains are scattered about. There are parts of leg bones, isolated lower jaws, and single hollow bones, some with old fractures; above all, the diversity of the prey shows how life went on there in that swampy bend of the river. Numerous remains of turtles and snakes give this impressive alligator feeding ground its special character.

Furthermore, feeding grounds and concentrations of remains of vertebrates that have been eaten are quite often found around dens and nests. To this category belong the often large accumulations of vertebrate remains at eyries and nests of both diurnal and nocturnal predatory birds. Ospreys, sea eagles, merlins, gulls, and other species of bird often carry the fish they have caught some distance away to an elevation, rocky knoll, or similar place where they then devour them. This is how scales and fishbones accumulate in rock crevices. Nehring (118) examined this phenomenon in detail because he found isolated remains of fish near Westeregeln [East Germany]. Wiman (176) cites the results of research carried out by the Danish natural scientist Lund. In Brazil, Lund found a small, vertical cave in which a pair of owls (*Strix perlata*) was nesting. Excavation showed that the whole cavity was filled with owl pellets, which contained the remains of roughly 7.5 million animals. Cave-dwelling carnivores cause formation of similar accumulations. But feeding grounds in open land are also fossilized, as Abel (7) has just recently described again.

In the oreodont beds of Nebraska, it is evident that predators found and devoured an abundant prey. The carcasses of oredonts have been torn apart and the parts carried off. The carrying off of parts that have been detached from a carcass naturally depends on their weight, and so a curious form of sorting takes place. The bones have been bitten into and then gnawed by rodents, as so often happens still today. Where the remains of the prey pile up, the predators' droppings are scattered everywhere. Carnivores often had the habit, shared by living hyenas, of forming bone depositories. We could not otherwise explain why we find skulls and lower jaw fragments of *Hoplophoneus, Hyaenodon,* oreodonts, *Leptomeryx, Perchoerus,* and *Mesohippus* all mixed together in one place.

6. PARTIAL BURIAL

Carcasses are often only partially buried. Osborn shows (124, p. 56) an illustration of a skeleton buried in aeolian deposits. It is one of the few examples in the literature of a recent carcass being compared with a fossil one. The upper part of the plate shows the partial burial of a young bull on the plains of South Dakota; beneath, for comparison, is the partially exposed skeleton of the fossil forest horse *Hypohippus* from the plains of Colorado. We have also found partially buried saurians in shelly limestone.

Another example of partial burial of a skeleton is found in pl. 14, fig. B. The cow carcass lies buried axially parallel to the shore of the Gulf of Mexico. When the water is high, waves can reach it, but when they do, the water flows around the obstacle and forms a semicircular depression in the shell-strewn beach. The weight of the original carcass and the formation of quicksand have caused it to sink into the sand up to its spine.[23] The hide of the exposed side has dried out like parchment, and underneath, the nimble sand crabs have completely skeletonized the carcass.

7. CARCASSES ON FACIES BOUNDARY LINES

Beds containing vertebrate remains are usually of great interest to those studying sedimentology and petrography. Burial often takes place near the shores of shallow bodies of water. In such beds, many signs provide information on the prevailing currents: ripple marks, plant remains, strandlines of organic material, clay galls, fish scales, animal tracks, and the like. Because a sizable carcass lying on fine-grained sediment presents a considerable obstacle to the movement of water, we can usually make very interesting studies, in the area surrounding it, of the prevailing currents and the origin of the sediments. A large obstacle restricts the flow of oncoming water, causing an increase in its velocity, which in turn means increased force; consequently, the water on the shoreward side of the obstacle can dig out a deep hollow, while sediment accumulates on the other side. This often causes the carcass to lie crooked or to sink in (pl. 14, fig. C, and pl. 27, fig. A). The fact that on the border between facies we observe sizable vertebrate remains aligned not only vertically but also, and most often, horizontally is related to this phenomenon.

Plate 12, figure D shows the carcass of a shark I photographed April 13, 1925, between the mouths of the Brazos and the Bernard rivers, near Freeport. The carcass lay on the borderline between two sediments: on

23. When only one side of a skeleton is embedded, the bones of the unprotected side are naturally often lost. When the mammoths of the Sanga Jurakh Creek [USSR] were excavated, it was seen that, with the exception of a few broken ribs, the protected parts dug out of the riverbed all belonged to the right side of the body. The left half had been exposed earlier and carried off (see Pfizenmayer, p. 225).

one side the deposit is coarse, consisting mostly of a shell bank (*Rangia cuneata*), while on the other there is only a sprinkling of shells, and the beginnings of sanding-over are evident. (*Rangia cuneata* is a brackish-water clam, which probably washes up in large masses on certain parts of the coast of the open gulf because brackish-water lagoons that have been destroyed are today being reworked under water.) This asymmetrical bedding on either side of the carcass is of special significance, and we must pay attention to it when we study fossil material. Notice that this cylindrical shark carcass is lying crooked—partially on its back. The side away from the main force of the water is buried deeper than is the other side.

The specimen shown in plate 14, figure A is also typical; it is the carcass of a ganoid fish over one meter long (*Lepisosteus*), which I photographed on the coast of the Gulf of Mexico between the mouths of the Calcasieu and Sabine rivers. A fairly coarse, wide beach ridge of shell fragments terminates landward in a steep, sloping embankment. Behind it is a channel that has been kept open by water flowing over the embankment and streaming back again over the margin of the beach. Here, a fine, muddy grain-size interfaces with the coarse, calcareous sediment because a sort of filtration of the water has taken place. On the other side of the channel, a disintegrating clay bank outcrops, and behind it is quartz sand. The fish was probably carried out of one of the river mouths and, following the pattern of beach filling, carried westward just as the shell fragments were and deposited on the beach by waves. We see clearly that its carcass divided the water that was carrying the shell fragments in such a way that its body caused an inlet to form in the pile of shells. The side on which it is lying is almost half buried in the shells, while the other is completely exposed. This example also shows how quickly such an accretion can form.

The position of the fish is reminiscent of many examples from the geologic past, where the sediment on one side of the carcass differs from that on the other, and there is perhaps even a third lying on top of these. I would like to present a fossil example from the Muschelkalk at Bennstedt [East Germany] in the Bruckdorf-Nietleben Muschelkalk syncline (see fig. 5). It is a small nothosaurid skull lying in a very thick layer of limestone completely devoid of fossils. Behind it, however, is a widening wedge of aphritelike shells and fragments of organisms, the accumulation of which is due to the obstruction of the waves. The same can be seen in the area of the foothills of the Harz Mountains [West Germany], where behind the shells of large *Arietites* lie accumulations of shells and shell fragments, pieces of wood, and other debris.

A great many vertebrate localities show changes in the vertical sequence of facies; an example is the *Mastodonsaurus* carcass assemblage at Kappel, described by Wepfer (173). The bones are often buried a few centimeters in the underlying sand and covered with clayey mud, guaranteeing good preservation. This bed of clay has also been partially reworked and fragmented. Moreover, the bones almost always lie under a covering one to two centimeters thick of almost pure clay, above which are layers of sandier clay. Some of the remains also lie completely in sandstone, or exactly at the interface between deposits of sandstone and sandy clay, or in a green sandstone deposit between an underlying sandstone, separated from it by a clay skin and an overlying red-green lean clay.

At the excavation conducted by the University of Tübingen near Trossingen [West Germany], isolated teeth and bones from Phytosauria and Saurischia are found throughout the whole thickness of the Stubensandstein, but partially articulated skeletal elements occur only at the upper limit where the sandsone interfaces with the gray-green marl. Huene (74) writes on this subject as follows:

> This marl does not form an even layer on the sandstone because the upper surface of the sandstone is uneven and in profile would present differences in elevation of from one to one and a half meters; one has the impression that the upper surface of the sandstone has been eroded here and there, and the marl deposited in the channels; at other places, sometimes very near by, interbedding of both rocks with sudden wedging out occurs; large sandstone nodules are also found in the lowest layer of marl. The marl lacks clear bedding, but is run through in all directions with slickensides, fault striations, and small slip faces. Accumulations of large grains of sand also occur in it, giving the impression that the marl was deposited quickly under water. However, these marls are by no means part of the red Knollenmergel of the upper Keuper, which overlies it some meters higher up. At the point we are talking about, the gray-green marl is at least four meters thick and does not come into contact with the Knollenmergel in this area.
>
> The saurischian skeletons are found only where the sandstone and marl are interbedded. The same skeleton may go from sandstone into marl and out again. There was a *Teratosaurus trossingensis* [*Plateosaurus engelhardti*] whose entire tail was in marl, but whose legs went down into the sandstone below. The *T. suevicus* [*Plateosaurus engelhardti*] described below was buried mostly in sandstone, pressed flat, with the abdomen underneath and the back upright; but the parts that lay higher, plus a few isolated parts, were in marl; a difference in the state of preservation in one rock as opposed to the other was not noticeable. The skeleton of *Sellosaurus hermannianus* was also buried in sandstone interstratified with marl, but was evenly surrounded on all sides by marl and came into contact with the sandstone only underneath.
>
> The upper layers of the sandstone must have still been soft and saturated with water when the deposits of marl suddenly began. The saurischian skeletons washed up at the same time as the onset of the marl deposition, obviously from a nearby shore. I assume that they

arrive there as cadavers, because some of the parts are still connected in a natural way, but entire extremities and other skeletal elements are missing. Although phytosaurian remains occur more often in pure sandstone than saurischian remains, no phytosaurian skeleton has yet been found in this horizon (marl interface). But because a crown from a tooth from a *Mystriosuchus,* present before the burial of the *Teratosaurus suevicus,* lay nearby, I conclude that the phytosaur fed on the saurischian carcass, which has gaps in the skeleton that seem to confirm it.

Just as interesting as the vertebrate-producing Keuper are the rich vertebrate beds of the Permian and Lower Triassic that we find especially in the Karoo formation of South Africa and in northern Russia. The Scottish occurrences are less important, the Australian more so, and the Indian most of all. Huene (in *Fortschritte der Geologie und Paläontologie,* no. 12) has just recently treated in detail the inclusion of this interesting vertebrate fauna in the rock series of the Karoo formation. Whereas most of the skeletons are no longer completely articulated, the pareiasaurs, which occur throughout the series of the *Tapinocephalus* zone—but only in colored clay-marl, not in sandstone—present a different appearance. They are always found as articulated skeletons, with the back upright. They seem, therefore, to have become stuck in softened ground; other remains found in coarse-grained rock were swept there by high water. That is definitely the case with the remains in Russia, where we have long known of a fauna from the western slope of the Urals. Amalitzky's (86) excavations in the area of the Little Dvina and Suchona rivers are indispensible to our knowledge of the subject. These finds in the red marl are associated with sand lenses, which owe their origin to intermittent rivers. Through cementation, the engulfing sand lenses have formed concretions. The remains were concentrated post-mortem and are mostly whole skeletons, which had no time to disintegrate before burial and the subsequent swift hardening of the concretions. Amalitzky's excavations yielded, in addition to a number of new species, ten pareiasaur skeletons; in three years he removed 65,370 kilograms of bones.

The partial preservation before burial of certain soft parts is of extraordinary significance in the preservation of fossil vertebrate skeletons, even though these parts usually disappeared after burial. Stromer (158) made the interesting observation on the vertebrates brought out of the diamond fields of German Southwest Africa [Namibia] by Kaiser (84) that the matrix, normally a hard, red stone, was gray and soft where it surrounded many skeletal elements, obviously due to the reducing effects of decaying soft parts. The discoloration was pronounced in the area of the large chewing muscles on rodent skulls. The discoloration of red stone by organic substances is enormous, and we should welcome the fact that F. Büchler (31) has recently done research on it. According to this author, the discolored zone is twenty-six thousand times larger than the once-present organic remains that caused it. If the iron pyrite is evenly distributed, a very small quantity is sufficient for uniform discoloration. The green color, therefore, is contingent on the ferrous oxide content and the distribution of unreduced ferric oxide still present in the green lean clay. Thus, lean green clay is formed from red by reducing processes carried out by hydrogen sulfide, which either comes from finely distributed iron sulfide or is released by the carcass itself.

Compared to bone assemblages that owe their origin to onetime catastrophic events, there are many, many localities whose bone-storing qualities have persisted over a long period of time. Filhol (47) still believed absolutely that in the old Tertiary phosphorite at Quercy lay a homogeneous fauna of the same age. However, the fauna is very diverse and is present in the upper half of the middle Eocene, the upper Eocene, and the lower Oligocene. Later Oligocene, Miocene, and Pliocene were never found. The beginning of the formation is marked by the presence of *Lophiodon lautricensis* Noulet and *Pachynolophus,* which belong to the Bartonian. The end is indicated by mammals, mollusks, and birds belonging to the Ludian, Sannoisian, and Stampian. The Aquitanian, on the other hand, is not represented.

Bartonian

Fauna from Robiac, Gard [France]:
Lophiodon lautricensis Noulet
Pachynolophus sp.

Lower Ludian

Fauna from Euzet-les-Bains [France] and Hordwell (Hampshire) [England]:
Palaeotherium curtum Cuv.
Adapis magnus Gerv.
Dichodon cuspidatus Owen
Lophiotherium cervulum Gerv.
Plagiolophus annectans Owen
Anchilophus dumassi Gerv.

Upper Ludian

Fauna from Montmartre [France]:
Palaeotherium magnum Cuv.
Plagiolophus minor Cuv.
Tapirulus hyracinus Gerv.
Anoplotherium commune Cuv.
Xiphodon gracile Cuv.
Pterodon dasyuroides Blainv.
Adapis parisiensis Cuv.

SANNOISIAN

Horizons at Celas and Ronzon [France]:
Plagiolophus fraasi Meyer
Anthracotherium alsaticum Cuv.
Gelocus communis Aym.
Ancodus cf. *velaunus* Aym.

STAMPIAN

Entelodon magnum Aym.
Anthracotherium magnum Cuv.
Aceratherium filholi Osborn
Cadurcotherium cayluxi Gerv.
Lophiomeryx chalaniati Pomel
Prodremotherium elongatum Filhol

The sudden disappearance of the Titanotheriidae from North America is striking. Doubtless it coincided with the change of climate in the central Great Plains, which resulted in a typical geological anastrophe. According to Abel (7), these creatures were adapted to a more humid climate and its corresponding vegetation. The overlying clay is aeolian, but the adjacent facies testify to a temporary paludification of the whole area, which soon gave way again to arid conditons. Floods left carcasses lying around for other animals to tear apart. In this connection, Abel (7, p. 344, fig. 228) shows an illustration of a skeleton of *Oreodon cobersoni* [*Merycoidodon culbertsoni*] that does not lie where it died, but where it was left by receding high water.

8. THE FORMATION OF FACETED REMAINS BY FLOWING WATER

The way in which a carcass that has been washed ashore is buried depends a lot on the ratio of the surface area of the body to its volume, just as when bodies are buried in other natural ways. Here too the following familiar law is valid:

$$\text{Rate of Sinking} = \frac{\text{Mass}}{\text{Resistance} \times \text{Internal Friction}}$$

A smooth, cylindrical carcass behaves completely differently from one that is disk shaped or has an irregular shape. The most irregular are the carcasses of flying creatures—pterosaurs and birds.

Let us look first at flattened carcasses. Here, the basal area is quite large and the position, therefore, fairly stable. Plate 14, figure C shows the carcass of a ray[24] near Mitchell's Cut at the outlet of Matagorda Bay. The ray

24. These sharp-toothed fish, predators on shellfish, have come down to us often as fossils, and their appearance in enormous schools still today often requires that

(Continued)

lies on the inner side of a beach wall, and there is radial runoff toward the rear. We see how the water flowing in this channel causes the carcass to lie at an angle, and how the tail bends in the direction of the flow. The blazing sun has already dried out the creature considerably, causing the shape of the cartilaginous skeleton to become more and more clearly molded into the upper surface of the body. Jaekel (81) shows a ray (*Urolophus crassicauda* [Blainville]) in his plate 5, figure 6; we see clearly how flowing water has pressed the tail against the body and has probably also influenced the somewhat irregular silhouette [see fig. 6]. On the other hand, *Platyrhina egertoni* de Zigno sp., (fig. 7), shown by Jaekel in plate 2 of the same work, shows only a slight curvature of the tail toward the right side of the body. Also, the angel shark described by Drevermann (43), *Squatina alifera* Münster, rests firmly, like a ray, on its wide pectoral fins. The tail bends around to the right because it was still moved by a weaker flow after the torso had become securely anchored. This specimen reconfirms the rule: The part of the body with the least basal area is the most movable.

Let us look further at the irregular carcasses of birds and pterosaurs. First, a recent example: On plate 15, figure A, we see an oddly contorted carcass of a young pelican, whose pouch has already ruptured and decomposed. I photographed it in January 1925, near Cameron, west of the place where the Calcasieu River opens into the Gulf of Mexico. Because we have already seen in the carcass of *Urolophus* that the more flexible parts press up against those that are more firmly anchored because they have a wider basal area, we recognize in this bird carcass the unusually strong influence of the asymmetrical onrush of water on the silhouette and on the arrangement of the individual elements. The torso was anchored first. One side of the carcass forms a solid front, interrupted only by the angle between the bill and the neck. The neck has a kink in it, and the leg has been thrown backward in the same direction as the neck and wings. We see the accumulation of shells in the flow shadow. The seaward side has been thoroughly flushed and is completely devoid of shells of any size. This unavoidable bending around of the individual parts—and we see how forceful it was in the position of the spine, visible in the pelvic area, which lies with the dorsal side down—causes large numbers of skeletal elements to lie parallel to each other and to form consistent angles with other parts lying near them. So the pelican's neck, humerus, and leg bend away from the body in one direction, and the torso, beak, and distal parts of the wings all go in another.

This observation prepares the way for understanding disarticulated skeletons whose individual elements are arranged axially parallel to each

---

protective measures be taken against them. They are known to invade San Francisco Bay, where they devour huge numbers of oysters. Because they crush the shells of their victims, their feeding habits add vast quantities of shell fragments to the sediments.

other. The best example is probably the *Pteranodon* shown by Wiman (177) in his plate 2 (see fig. 8). Most of the bones shown in the drawing are lying parallel to each other. Although the upper jaw and the lower, lying beneath it, are pointing in opposite directions, they have not been separated but merely opened out at the hinge. Five other bones, not including some of the smaller ones, are aligned in the same general direction. A few other bones are lying at acute angles to this group. It is a typical example of potential water turbulence automatically rearranging the skeletal elements of a carcass that has become disarticulated due to putrefaction. The same author shows in plate 3, figure 9 [see fig. 9] of the same work a good example of the flushing of the front formed by an irregularly shaped, complete pterosaur skeleton (*Dorygnathus banthensis*), clearly lying on the bedding plane. When the already severely macerated carcass sank to the bottom, the torso became anchored, and the tail, wings, and head were moved laterally by water currents. In the process, the left femur came off and remained hanging between the teeth. A few vertebrae seem to have been carried farther off.

Broili (29) shows a pterodactyl that seems to me to have an obvious flow deposit on its left side, underneath its considerably displaced bones. The one he described that had remains of wing webbing shows both extremities on the left side of the body bent in a normal relaxed position, while on the other side, the bones of the fore- and hind limbs form a unified front at an acute angle with the axis of the body.

In the severely disarticulated skeleton of *Pterodactylus kochi* Wagner [Germanodactylus cristatus (Wiman)] in the Munich collection, Plieninger (129) has analyzed a similar situation. The animal was obviously buried while lying on its back. The skull was wedged somewhat to the side, and the bones of the limbs were all more or less thrown together to one side of the still articulated neck and back sections.

We can perhaps explain the position of the head of the *Platecarpus coryphaeus* described by Wiman (177) by saying that it, too, is a flow deposit (see fig. 10). Because of the severe curvature of the neck, the laterally lying head is bent around parallel to the right forelimb. The resultant pinching in the spine caused its double curvature.

9. THE LAW OF THE RIBS

When the vertebrate spine is bent into various lateral, ventral, and dorsal positions, the position of the ribs is governed by the curvature; the correlation is the same as for any movable appendage attached to a curved axis. When the body is curved, the ribs on the convex side are spread apart, and their free ends diverge, while the ends of those on the concave side converge. If, as usually happens, the body does not lie in a completely dorsal or ventral position, but in a more asymmetrical one, then the ribs on one side of the body bend at more of an angle than those on the other.

The ribs naturally tend somewhat toward the rear; those on one side, then, may lie at a very acute angle toward the rear, pressing against the spine, while those on the other side are splayed. I have often observed in conifer twigs from the Kupferschiefer that the needles on one side formed an angle different from that on the other side. We also find the same phenomenon in fish, for example, in the *Ameiurus primaevus* [*Amia uintaensis*] (related to recent silurids) from the Eocene of the Green River Formation of Wyoming, illustrated by Eastman (44a). On the left side of the body, the ribs lie at an acute angle toward the rear, but on the right side they are splayed almost at right angles.

If the long axis has been kinked, dislocated, or sharply curved, some really odd positions are seen. Figures 11–15 show us ichthyosaurs in uncomplicated positions; figure 16, on the other hand, shows the ribs on one side of the body divided into two groups. Even more remarkable is the female *Ichthyosaurus crassicostatus* Fraas [*Stenopterygius crassicostatus* Fraas] in the Senckenberg Museum in Frankfurt (fig. 17). Due to the raising of the head and shoulder girdle, the spine has become compressed where it passes through the rib cage. The convex sides of the forward ribs point hindward, and the opposite is true for the hindward ribs, so that the ribs on either side of the dividing line turn over as far as possible, causing their distal ends to diverge widely. We could call this an inversion of the ribs at the point of spinal compression. If, when the carcass collapses, the ribs lie flat on the ground, they are easily detached. In animals that have sunk into mud or broken through ice, the free head of the ribs can turn over and end up farther from the spine than the end fixed in the mud.

10. DESICCATION OF THE CARCASS

All the processes of growth and decay can be subsumed in the terms aggregation and dispersal. The cells and their membranes, blood serum and the sap of plants—all living matter is composed of colloids, and one of the most typical is protein (see Ditmar 42). In structural properties, a colloid has characteristics of both a solid and a liquid. There is an essential colloidal-chemical difference between living and dead tissue. When muscle tissue not yet affected by rigor mortis is placed in water or a salt solution, the quantity of water absorbed would increase quickly until, after twenty to thirty minutes, it reached a maximum, and then would decrease to its former level and even below that point. Not only the absorbed water, but also some of the water originally contained in the tissue, is given up. If pieces of kidney, spleen, or liver tissue are used in such an experiment, the result is the same. If a gelatin cube is used, a curve representing water absorption never falls. If a muscle is tested after rigor mortis has disappeared, it becomes apparent that it has completely lost its capacity for absorption and that it gradually gives up the water it still contains. The

capacity for absorption is therefore a special characteristic of living pro-
toplasm. When the body dies, the colloids decompose.[25]

As a carcass dries out, it shrivels, because when water is lost, two or
more interconnected layers of cells or tissue contract sharply and un-
evenly; at the same time, because the plasma remains attached to the cell
wall, cohesion mechanisms are also in effect. Therefore, with progressive
water loss, the cell membrane invaginates in places where it is thin, and
the size of the cell decreases. When a carcass dries out, these shrinking
mechanisms often bring about unnatural, contorted positions not usually
found in living animals. The most conspicuous is the often observed curvature of the cervical part of the spinal column.

Plate 13, figure A shows us the characteristic appearance of such a
carcass—a cow I photographed on a sandbank of the Brazos on March 21,
1925. The animal had originally drifted as a water carcass and was left on
the sandbank as the water receded. The tail end points downstream. The
pectoral girdle probably anchored first and served as the pivotal point
around which the carcass rotated. The skeleton is already quite free of
flesh, but some muscle fibers, ligaments, and tendons must be still preserved because the skeletal elements are still firmly articulated. Due to
drying out of the neck muscles, eaten later by vultures, and shriveling of
the large tendons, the neck has bent perpendicularly upward, so that the
rear part of the epistropheus [or axis] touches the neural spine of the first
vertebra of the trunk, and the head has come to lie far backward. It seems
that only the large spinal processes kept the head from bending back
even farther.

We can very often see this high-arched neck in fossils (see fig. 18). The
skeleton of *Palaeotherium minor,* shown by Cuvier (34) in his plate 34 and
described on page 214, is an excellent example; it comes from Pantin
[France]. Another is the *Palaeotherium* skeleton from the gypsum of Montmartre exhibited in the stairwell of the Museum of Natural History at the
Jardin des Plantes [Paris]. An outstanding example from Germany is the
specimen of *Diceroceras furcatus* [*Euprox furcatus*] from the Miocene at
Steinheim, shown by Oskar Fraas (49) in his plate 11. In plate 13, figure B,
I show a photograph of the original in the natural science collection at

25. Fürth attributed normal muscle contraction to phenomena of absorption and
release of fluid. According to him, muscle relaxation was due to the release of fluid
by swollen protein particles. Lenk and Fürth ascertained after numerous experiments that rigor mortis was not a process of coagulation but of absorption.
After death and the cessation of normal blood circulation, lactic acid is formed,
which causes the muscle fibers to absorb sarcoplasma. This phenomenon induces
rigor mortis. The release of fluid as a prelude to the draining of the colloidal system leads to the diminution of rigor mortis because the muscle protein gradually
coagulates as acid continues to form. Rigor mortis passes more quickly in the summer than in the winter, a fact that ties in well with the idea that lactic acid is its
cause. Oxygen hinders the onset of rigor mortis by destroying lactic acid. According to Lenk, a disturbance in the balance of fluids means death.

Stuttgart (with kind permission of its curator, Dr. Berckhemer). Here, the skull lies in the dorsal position and the lower jaw has been pressed to one side; this could easily have happened after the animal was buried. A skeleton of *Palaeotherium magnum* Cuvier from the uppermost Eocene (Ludien) from Mormoiron, Vaucluse, France, a drawing of which I reprint here (fig. 18), from Abel (5), shows a much more pronounced curvature of the cervical part of the spine. The skull is completely bent around so that its dorsal side is right up against the lumbar vertebrae. In the Eocene gypsum beds of the Paris Basin, this pronounced curvature of the neck is the rule. Another example of this position, from the realm of reptiles, is the well-known skeleton of *Compsognathus longipes* Wagner (167) from the Upper Jurassic at Jachenhausen in the Oberpfalz [West Germany]. The skull and neck of this animal are also bent around until they lie up against the pelvis (see pl. 37, fig. A).

The degree of curvature of the neck depends to some extent on the development of an anticlinal vertebra, which Gottlieb (59), using recent information, has treated in detail. In many mammals, the spinal processes of the vertebrae of the trunk increase in size in both directions from a vertebra [the anticlinal] lying somewhere in the middle of the spinal column. In the anterior part of the column, the bodies of the trunk vertebrae are inclined forward, and typically, the smallest vertebra also has the least inclination. In *Echidna* [*Tachyglossus*], all the processes tend slightly and evenly toward the rear. Among marsupials, the development of the anticlinal vertebra varies: In the climbers it is not in evidence; in many hopping forms there is a slight anterior tendency of the processes in the lumbar area; the insectivores have in many cases already acquired the anticlinal vertebra. In bats it is only hinted at, and only in the large forms; in rodents, it is well developed. The form of the spinal processes of swimming creatures is different and, in those with no teeth, shows multiple structures for reinforcement.

Of the carnivores, the cats always have a vertebra that can be called anticlinal; in bears, however, it is totally lacking. In seals, all the neural spines are alike, but in walrus, they are not. Only whales with a sharply arched vertebral column have the anticlinal vertebra. It is often lacking in ungulates; manatees have none. Among the primates it is clearly formed in the prosimians, but among the simians it is found only in tree dwellers.

When an animal dies, the anticlinal vertebra, the presence of which entails a change in direction of the spinal processes, has a decisive influence on the curvature of the spine and thus on the position of the body. This type of structure is found in mammals and, like a suspension bridge, it bears the weight of the entrails and supports itself on the extremities. The resistance to shock must be fairly great. Resistance to bending of the spine, achieved by means of the long spinal processes lying close together, comes at the expense of mobility, as the large ungulates show (pl. 13, fig. A). But even in the forms that move with agility, reinforcement of the dorsal vertebrae is especially important. What is decisive for the development of an anticlinal vertebra are those muscles that insert upon the spinal processes and operate in cooperation with the limbs, and carry the trunk or control the bending of the spine (cucullaris, latissimus dorsi, multifidus, levator caudae, rhomboideus). The body's center of gravity lies in the area of the anticlinal vertebra. When the back is overloaded, further sagging is impeded by the spinal processes coming into contact with one another, facilitated by the opposing directions of the anterior and posterior processes.

The weaker the spine is, the more pronounced is the curvature of the neck due to differential drying out of the tough dorsal parts and the softer, more easily decayed viscera. Plate 12, figure B shows the mummified carcass of a soft-shelled turtle, photographed March 15, 1925. The long neck is curved back into a pronounced S shape, causing the head to lie on the right side of the carapace. The legs are passively extended, as I have always observed in dead turtles. Very noteworthy is the pronounced shriveling of the membraneous part of the shell.

Most characteristic are the curvatures in the necks of long-necked birds. Best known is probably the pronounced backward bending of the neck of *Archaeopteryx macrura* Dames [*A. lithographica* von Meyer] from Solnhofen. (Reproductions of this valuable treasure of the Museum of Natural History [Museum für Naturkunde], in [East] Berlin, are found everywhere, and it is worthwhile to compare the position of this creature with that of the *Compsognathus longipes* Wagner.) In general, the best examples are provided by the pterosaurs (see pl. 34, fig. C), as long as they were not just water carcasses but were subjected to a more thorough drying out, as was the often-pictured carcass of *Pterodactylus longirostris* [*Pterodactylus antiquus*], whose position we can compare directly with that of the *Compsognathus* and the *Archaeopteryx*. Also, the specimen shown by Meyer (110) in plate 1 shows the characteristic curvature, and the specimen of *Pterodactylus elegans* Wagner, which Zittel (185, p. 355, fig. 77) has shown, is similar to these examples.

Heinroth (65) has investigated the backward bending of the neck after death in *Archaeopteryx*. He chose two species of birds that resemble it in outward appearance: a magpie and an Australian pheasant cucal. He arranged them in the position of the *Archaeopteryx* and took interesting photographs of them fresh, plucked, and finally, unfleshed. He ascertained that after the detachment of the muscles and some slight decay, the heads of both birds fell backward, exactly as we see in the Berlin *Archaeopteryx*. What happens is that after muscle tension has disappeared, the pull of the ligaments comes into play; according to Heinroth, the sleep position of many birds also reflects this process.

In connection with this backward bending, we must also note that in rigor mortis, the antagonistic muscles all contract at the same time.

Moodie (115) tried to explain the process resulting in this curvature as being pathological. He saw in the backward bending of the necks of numerous long-necked fossil skeletons such as pterosaurs and *Compsognathus* a kind of tetanus or opisthotonus. Huene (76) correctly countered this argument by pointing out how often such a position occurs, and that it comes about after death, not before, and therefore cannot have anything to do with disease. Von Huene's observations, made on his journey across the Patagonian pampas, are very interesting. He came across the carcasses of numerous guanacos and rheas, all of which showed this pronounced curvature of the spine. The cause was the drying out and contraction of the musculature after death, appropriate in the dry climate. Dienst, too, came to the same conclusions. He gave me permission to use the photograph reproduced on plate 13, figure C of a dead camel he saw on the way from Aswan to the oasis of Curcur [Egypt] in March 1909. Here, too, the neck has dried out and curved dorsally, so that the base of the skull lies almost on the trunk, and the head, analogous to fossil skeletons, points perpendicularly upward.

The preservation of *Archaeopteryx* and pterosaurs with backward-pointing heads also caused a small controversy between Deecke (39) and Hennig (66). The former was of the opinion that these animals got stuck in the mud while diving in shallow water for crustaceans. Hennig pointed out that the tails of two East African stegosaurs also lay curved in a half circle, and that *Diplodocus* presented a similar position. He also pointed to the dorsal curvature of *Leptolepis* from Solnhofen, of *Pholidophorus pusillus* from the alpine Triassic shale, of *Eurylepis tuberculatus* [*Parahaplolepis tuberculata*] from the black shaley coal of the Carboniferous [Pennsylvanian] in Ohio, and of *Rhinellus furcatus* from the Upper Cretaceous of Westphalia and Lebanon. Hennig formulated the following generalization: "Dorsal backward curvature is not an indication that water-dwelling vertebrates died on land, or that an aerial form made an unlucky dive, but a phenomenon that occurred automatically after death and before actual burial took place, uninfluenced by external circumstances and independent of way of life or manner of death."

Plate 12, figure C shows a photograph of a characteristic example of dorsal spinal curvature taken at the carcass assemblage at Smithers Lake in March 1925. The pectoral girdle, sternum, fore- and hind limbs of a wild duck have completely rotted away; the skull and sacrum, however, are still connected to the spine, which bends at an acute angle almost exactly in the middle, so that the skull and sacrum are extremely close together. The head lies to the side, and the sacrum is semidorsal, producing a slight twisting in the axis of the spine. The lower jaw is splayed in a way that has become familiar in many pterosaurs. In terrestrial sediments, slightly curved skeletons such as that of *Sclerosaurus armatus* H. von Meyer (pl. 37, fig. D) are often found.

11. HOOKED, BENT, AND CURVED CARCASSES

If one visualizes a drowned, long-tailed pterosaur floating in shallow water, one can easily see that the skin of the wing would offer resistance to moving water, putting a strain on the pectoral girdle. Were such a carcass beached, the head and pectoral girdle would probably anchor in the substratum first, with the tail and wingtips pointing toward the shoreline. If the water level were to rise, the carcass would be flushed by retreating waves. In the process, the anchored head would not alter its position significantly, but the lightly floating tail and the tips of the wings would probably be folded back; the resulting curvature, reaching its maximum in the neck or the area of the pectoral girdle, would cause the trunk and head to lie parallel to each other or even to touch. Should this happen, the wingtips and the pectoral girdle would be easily detached altogether (as we see in an actual specimen in pl. 34, fig. B) and thrown parallel to the spine, on top of each other but in mirror image, the ends diverging. On the other hand, the specimen in plate 36, figures C and E, has no limbs at all; consequently, we also find isolated forelimbs (pl. 36, fig. D), which anchor more easily when part of the pectoral girdle is attached and behave just as a whole pterosaur skeleton does. The effect of anchoring is the same as if the whole, flexible spine had been left hanging on a post stuck in the sand, leaving the rest of the carcass floating in the water, to be anchored later according to the flow around the post.

Plate 36, figure C shows the backward-turned skull and lower jaw exposed from the right side. The jaws are opened wide.

The skeleton in plate 34, figure B was found in 1857. The skull and the whole spine are present. Notice particularly that, contrary to what we see in plate 36, figures C and E, both forelegs are still relatively naturally articulated. Nevertheless, we see how under the effect of the flow, the pectoral girdle and the forelegs are thrown parallel to the body, one underneath the other, and are separating from the spine.

The skeleton in plate 36, figure E is also missing its forelimbs, but the skull, the whole spine, the breastbone, and the pelvis are present. The skull is seen from above. Just behind it, the cervical vertebrae are in their natural articulation. The rest of the spine, somewhat compressed, follows as it should, and the pelvis is still attached normally. Only the breastbone has been torn loose and bent fairly far backward. In the so-called Haeberlein specimen, the lower jaw is missing entirely. Right behind the back of the head, only a little off center, lie seven cervical vertebrae in an unbroken line. Then there is a small space, after which the rest of the vertebrae follow without interruption. Although the cervical vertebrae and the dorsal ones just mentioned are lying laterally, the rest are lying ventrally.

This position, which recurs so often, is easily explained. If one holds a bird carcass that has not yet been affected by rigor mortis by one wing and lets it dangle, the same hooking appears, with maximum curvature at the base of the neck on the side turned toward the ground. To transpose this operation to the horizontal, let us imagine there is a small stake planted in shallow water where a carcass floats. It would form an obstacle, and the currents would fling the carcass around it so that the head and tail or hind limbs pointed shoreward. If we replace the stake by the anchoring of the carcass, the maximum curvature now lies not at the middle of the spine, but at the center of gravity. An obstacle lying on the beach divides and dams up the onrushing or receding water; the water pushes to the sides, is laterally constricted, and increases in force proportionately. The parts on the inner side of a curved carcass are, however, somewhat protected.

These observations are based on the clear-cut example of pterosaurs. The carcasses of birds (pl. 15, fig. C) and of awkwardly shaped reptiles and mammals assume exactly the same kind of position.

Pterosaur skeletons differ from each other considerably in degree of disarticulation, which depends both on the intensity of the movement of the water and, more important, on the length of time the carcass is exposed to it. The extremities with the long wing-fingers and webbing are especially vulnerable to stress and are carried off; the breastbone separates; the lower jaw comes off; the ribs become dislocated.

The hooked position in pterosaurs just described is also found often in long fish. The illustrations in plate 15, figures B and D show two specimens of *Aspidorhynchus acutirostris* Blainville from the collection of the Geological Institute of the University of Halle [East Germany]. These are large individuals that show many characteristics of the typical disarticulation of ganoid skeletons: dissolution of the attachments in the skin that hold the scales, curvature of the spine, and disconnection of the parts of the skull separating from the parasphenoid. The first specimen (pl. 15, fig. B) is completely hooked. The entire spine, however, has been torn out of the scaly skin, and from the tail fin on does not follow the curve of the rest of the body but diverges from its axis at an acute angle and extends right up under the middle of the skull. It is possible that with increasing putrefaction, the water turbulence brought about this result; I think, however, such a situation is best explained as having been caused by the feeding of another animal. For the best comparison, see the pike carcass from the flooding of the Elbe (pl. 7, fig. B), where the fish's spine has been pulled out and disarranged by feeding crows. Obvious examples of the tearing out of spines by scavengers are also found in a number of Solnhofen fish in the Staatssammlung at Munich.

In the second example (pl. 15, fig. D) a very curious observation can be made. Originally, this carcass had also assumed a simple hooked shape; then, however, due to some external influence, the spine and part of the scaly skin of the hind third of the animal were pulled out in the direction of the inner curvature. Where the body was concave, it is now convex. Oddly enough, the place where the spine originally lay in normal association can still be clearly seen. A spindle-shaped surface is enclosed on one side by a furrow representing the carcass's original position and on the other by the carcass itself. Here, due to feeding by a scavenger, we see the beginning of a process that is much further along in the previous example. Even though I posit external intervention, we must remember that the spine within its curved envelope of scaly skin is subjected to tension and, because it is elastic, seeks and eventually finds release when the ventral side bursts. After the flesh decays, the spine of the ganoid lies considerably displaced within the scaly hide.

In connection with the hooking of firmly anchored carcasses, it frequently happens that the spine bends at an acute angle and breaks because it is subjected to more stress than it can bear. Plieninger (128) described a specimen of pterosaur, *Campylognathus zitteli* [*Campylognathoides zitteli*], from Holzmaden, that shows the phenomenon very clearly. The spine in the area of the pectoral girdle is bent back at an acute angle, causing the dorsal side of the skull to face the inside of the angle. Outside the angle, the lower jaw lies parallel to the skull; within it, the pelvis is parallel to the rest of the spine. Another example is even more extreme: Nopsca (121) described the new genus *Eidolosaurus*, from the Neocomian, and had this to say about the reptile's position:

As the cast shows, the anterior part of the creature lies mostly on its back; the abdomen, therefore, is turned toward the observer. However, the unusually large tail, separated from the anterior part, lies on its side. Behind a skull fragment, the cervical and thoracic vertebrae form an undisturbed row. There seems to be a break in the area of the scapula, and we will see later why this is so. The ribs on the left side are in the natural position. On the right side, the ventral ends of the anterior ribs are together and point toward the skull; the posterior ribs on that side retain their natural position.

The only major disturbance in the skeleton of our creature occurs in front of the sacrum, where the natural articulation of the spine is interrupted. The sacrum and the tail are turned around roughly 160 degrees. The distal end of the tail lies across the front extremities and points almost toward the cranium. The hind legs are both firmly connected to the pelvis and lie at a slant facing posteriorly on either side of the sacrum.

The disturbance in the lumbar region requires a comment because it is also noticeable in almost all other Lacertilia known from Istria [Yugoslavia]. In *Actaeosaurus*, a dislocation is noticeable; in the specimen of *Adriosaurus* in London, there is a slight dislocation in front of the sacrum and a pronounced one behind the supposed area of the anus; in the Vienna specimen the only dislocation, albeit a pronounced one, lies right behind the sacrum. In *Pontosaurus*, there is a slight disturbance of the spine both in front of the sacrum and in the area of the anus.

Similar dislocations, obviously facilitated by putrefactive gases in the intestine, as in the dolichosaurs, are also noticeable, although to a lesser extent, in the aigialosaurs; however, the *Dolichosaurus* always differs from the *Aigialosaurus* in the position of the tail, which, in the former, is frequently folded back. We might see this disturbance as having been occasioned by predators feeding on the carcass, but there are factors that indicate another explanation.

In most of the dolichosaurs, almost all the other bones of the skeleton lie undisturbed; furthermore, in our new remains, along with the tail disturbance, we see a shift in position of several ribs of the right side; they have been moved away from the body and point in the same direction as the tail. This repositioning of the ribs shows that the same force that moved the tail affected the whole decaying carcass, and that the concave side of the body naturally offered a better attack surface than the convex. Instead of thinking of a local attack by carnivores, we must think of a more generalized force at work. As the long hemapophyses of our new remains show, the creature had a long, wide, laterally compressed tail, which naturally offered a large contact surface to every current of water; therefore, as putrefaction set in, displacement of the hind part of the body could easily occur. What is special about our remains is that this displacement of the tail must have been the consequence of a current moving obliquely toward the cranium and against the ribs. Because the tail displacement is never present in the aigialosaurs, we can conclude that they did not have a wide, laterally compressed tail.

Therefore Nopsca, too, thinks in terms of stationary currents. Observations of this kind are also valuable for estimating the depth of the water and the genesis of the burial medium. Long fish carcasses often show this phenomenon of the acute-angled bend; indeed, the fish of the Glarner Shale are the best-known examples. Wettstein (174) shows a number of fishes reflecting this phenomenon. His illustrations pertinent to this subject are the following:

1. *Lepidopus glaronensis* is bent twice at right angles, so that the head is once more parallel to the axis of the body (pl. 5, fig. 5)
2. With two bends (pl. 5, fig. 4)
3. *Lepidopus brevicauda* bent at an acute angle (pl. 5, fig. 7)
4. *Lepidopus glaronensis* bent at an acute angle (pl. 5, fig. 9)
5. *Lepidopus glaronensis*—the spine has been torn out so far that the separated parts are perpendicular to each other; moreover, it is bent once at a right angle and once at an acute angle (pl. 5, fig. 10)
6. *Lepidopus glaronensis,* bent at an acute angle, an obtuse angle, and

repositioned to form a right angle with the curved spine (pl. 6, figs. 1, 2, 3; see fig. 19)

7. *Lepidopus brevicauda,* bent at a very acute angle (pl. 6, fig. 4)
8. *Lepidopus glaronensis,* extremely fragmented (pl. 5, figs. 5 and 6)
9. *Lepidopus glaronensis,* bent at a right angle (pl. 6, figs. 7, 8)

As early as Agassiz (10, pls. 37 and 37a of vol. 5), we see illustrations showing bent fish skeletons from the Glarner Shale.

Indeed, in every large collection, documentary evidence of this phenomenon is found. Marck (105) published a study of a fairly long fish from the Cretaceous limestone, *Leptotrachelus armatus,* whose anterior end is bent at an acute angle. Several times, I observed a similar bending with overlapping of both parts in long pieces of fossil wood in the Mansfeld Kupferschiefer. A good example is found in the collection of the Bergschule at Eisleben. Sometimes the fish of the Kupferschiefer also show this bending already described, but because their bodies are thickset and not so long, their appearance is a little different. Freygang (54, fig. 26) made a schematic drawing of the position of the Kupferschiefer fish. He does not attribute the bend to a current, but assumes that the fish carcass floated straight down through the water, touched bottom first with its head, finally buckled, and was covered up. On plate 15, figure E, I give another example. It is a *Pygopterus humboldti* from the Pangert collection at Eisleben, completely bent out of shape. The head, readily recognizable from the teeth, touches the tail fin. Because this creature is unusually sharply hooked, the pectoral, ventral, and caudal fins are spread out at sharp angles to the periphery. At the maximum curvature in the middle of the body, a characteristic bulge shaped like an hourglass has formed due to overlapping. Weak currents predominated at the time of the formation of the Kupferschiefer.

We have found whole schools of fish lying on the larger slabs from the old, shallow mineshaft. I know of a number of plant remains from the Kupferschiefer that lay axially parallel to fish. There is an example in figure 20—a *Taeniopteris* leaf and a *Palaeoniscum* lying side by side. Oddly enough, the base of the leaf bends in the same direction as the fish. Another specimen shows remains of *Baiera digitata* and *Palaeoniscum.* In the Wohlbier collection, there is a thin piece of wood lying with a *Palaeoniscum* in the dorsal position. Quite often there is bare rock lying between remains that exhibit the same orientation. A further instance also involved *Baiera* and *Palaeoniscum.* Fragments of conifer twigs lying axially parallel to each other also occur. Naturally, there are also other arrangements, but certain currents can be inferred that have probably been transposed from the surface water to the depths.

The positions of the hooked fish also conform to each other. Two hooked fish lie absolutely symmetrically arranged, about a handbreadth apart, on a slab in the collection of the Bergschule, at Eisleben. The parts with maximum convex curvature turn toward each other, and the head and tail ends diverge proportionately. On another slab, in the collection of the Geological Institute, at Halle, are two curved fish also arranged symmetrically; the tails overlap and the heads diverge. Another slab, also in the collection of the Bergschule, at Eisleben, differs in that the fish are head to tail, and the opposing bent ends of the bodies diverge symmetrically. In such instances, the attraction that swimming bodies exercise on each other has been preserved in the fossil record after their simultaneous death.

In spite of many analogs of facies and genesis, if we compare the burial of vertebrate remains in both the Kupferschiefer and the Posidonia Shale, we come to see certain differences. In the Kupferschiefer, even though there are indications that turbulence was slight, there are fish lying curved in a circle, and bent and folded ganoids are present. In the collection of the Geological Institute in Greifswald [East Germany], there is a large *Acrolepis asper* whose torso, curled around some sediment, forms such a complete circle that the head and tail touch. All the saurians, not just those described by H. von Meyer (108), but also the more recent finds such as the new one at Wolfsschacht (pl. 17, fig. B—in the Göttingen collection, and not a *Protorosaurus*), lie with the most pronounced curvature. The most interesting is the skeleton of a small saurian from the Mansfeldite, *"Aphelosaurus"* [*Weigeltisaurus*], studied by Jaekel, and called a pterosaur when first found. The creature lies peculiarly compressed underneath the fin of a large *Coelacanthus,* with its head perpendicular to the axis of the sheltering fin and the widely splayed forelegs parallel to the edge of the fin. The spine has two right-angle bends, causing the midsection to go in the same direction as the head, and the other two parts, both pointing to the same side, are parallel to the rays of the fish's fin.

In the Posidonia Shale, however, circular skeletons are only rarely found in the Jurassic. The stratification of remains is seldom influenced by water turbulence, and the preserved outline of the body often reflects undisturbed burial without previous drying out. On the other hand, in the Kupferschiefer, severe drying out of the adjacent areas is documented, and the extremely curved carcasses were later transported from these areas by water and buried with other remains that do not show the phenomenon. The water was fairly deep and still, although not totally devoid of currents. Even the bones of completely disarticulated ichthyosaur skeletons lie together in such a way that only scavengers can account for the displacement. The number of severely disarticulated skeletons is considerable. Only the bands of *Coeloceras* shells and the positions of certain pterosaurs with flow deposits indicate that during severe storms there was some effect, however slight, from water turbulence.

Based on the way shell remains are interstratified, Berz (22) has recently deduced similar water turbulence, as I (172) have done for the Posidonia Shale of the Lower Carboniferous of the Harz Mountains [East Germany]. He presents as evidence the layered accumulation of fossils and deposits of shell debris. Nevertheless, it seems that periods of calm alternated with stormier times, and only during the latter would the deeper water have been affected.

According to F. von Huene (75), the remains of ichthyosaurs are always found alone and fairly far apart. The young, well-preserved individuals "are found stretched out in a completely undisturbed position, mostly lying on their sides."

In many specimens of *Stenopterygius crassicostatus,* the front of the torso lay almost perpendicular to the head and above it, and from the middle of the torso on, the skeleton was lying on its side.

Hoffmann tried to verify the position of the Mansfeld fish experimentally, and Wöhlbier (181) reported on the results as follows:

> I cannot go into great detail here on Hoffmann's further research, but I would like to make the following brief statement:
> Most people think that most fish found in the curved position died a quick, violent death. But because of this very curvature, it seemed likely to Hoffmann that they died a natural death, sank very gradually, and little by little were covered by sediments drifting down through the water. Hoffmann performed the following experiment: He placed a perch he had just killed in a container of water into which he had thrown a handful of sand. For four days, the fish remained stiff and stretched out on the surface of the water. Then it gradually became limp, and the head and the tail began to curve downward. For twelve days, it held the exact form of one of the curved fish found in the shale. Because the back was heavier than the abdomen (the body collapsed because the innards had already decayed), the fish lay on its side. On the fifteenth day, it sank down and hit the sand tail first. In the days following, the head, too, finally came to rest on the sand. On the eighteenth day, the fish lay on its back, and the tail gradually spread out over the sand just like the tails of the Mansfeld Shale fish.

Freygang (54), too, assumed that a fish curves after death and floats in the water on its side with head and tail pointing downward. As it sinks, head and back hang down at a slant, while the abdomen, full of putrefactive gases, points upward. So the fish often comes to lie on its back with the tail curved laterally. If the carcass turns and falls on one side, the curvature disappears again; if it remains upright, it finally bends. Almost all curved fish show the midline of the back with rows of scales going out from it at an acute angle. The dorsal position is frequent, but sometimes the ventral position occurs. The recent ganoids from Smithers Lake offer many analogs. The dorsal position is also frequently encountered in saurians from other localities. The *Proneusticosaurus madelungi* described by Volz (166) lies on its back, with the toes of the fore- and hind legs touching.

The position of the Kupferschiefer fishes is also often found in ganoid fishes. Pander (125, fig. 8, pl. 2) shows it in *Osteolepis macrolepidotus* from the geodes of Lethen Bar [Scotland]. The *Amblypterus* from the Rotliegend (Langenhahn 7, pl. 1e, fig. 4a, and pl. 3, fig. 1) shows it, and so does the *Rhabdolepis cretaceus* from the platy limestone of the Upper Cretaceous (Marck 105).

The unusual position in which the dorsal midline occurs where one might suppose the "lateral line" to be seems completely natural when compared with recent ganoid fishes and is also found in a number of fish localities, especially in the Devonian. Newberry (119a) came upon three examples of this phenomenon, clearly seen in his illustrations of *Catopterus ornatus* Newb. (pl. 18, figs. 3, 3a, 3b) and *Catopterus minor* (pl. 17, figs. 3 and 4) [both now referred to as *Redfieldius gracilis*]. He writes as follows:

> The body must have been round or somewhat flattened vertically, since it lies on the abdomen with the middle line of the back uppermost, the position generally assumed by the fishes which I have designated by the name of *C. minor*. The general aspect of these fishes is so similar, that I have been inclined to consider them as varieties of the same species.... From Durham, Conn., I have obtained through Mr. Loper quite a number of small specimens of *Catopterus*, which are of nearly uniform size—about five inches in length by one and a quarter in width—all lying partly upon the abdomen and showing the median line of scales upon the back. This proves that the body was round or perhaps somewhat flattened vertically. The head is small, depressed, pointed.

As for the ventral position of *Palaeoniscum freieslebeni*, really not so unnatural for ganoid fishes, Germar posited a triangular cross section with the flattened abdomen as the base and the back forming the apex. In many Kupferschiefer fish, the upper side is well preserved, but the part on which it rests has become completely disarticulated, and scales are strewn about, forming an aureole around the remains (pl. 17, fig. C); the opposite position also occurs. Isolated shreds of scaly skin are also found (fig. 21), as are strips with only a single row of scales. The large ganoids from Solnhofen often provide outstanding examples of the latter situation (see pl. 15, figs. B and D).

Based on shriveling of the abdominal areas of fish, Lambe (96) concluded that the environment of the "Albert Shales" had dried out from time to time. With great care, Pompeckj (130) applied this line of thought to the Kupferschiefer. He is completely right when he maintains that the findings in the Kupferschiefer are different. And yet, desiccation cracks occur unusually often in the immediately subjacent white beds at the base of the *Flözberge* and are easily seen at the Wolfsschacht, and I assume that the margins did in fact dry out; in addition to water carcasses, we also

find mummified carcasses that had been lifted out by rising water and reburied.

There is one position observed elsewhere, especially in Jurassic fishes, that seldom occurs in the Kupferschiefer: the one in which the head is bent upward laterally, the back is concave, and the abdomen convex. Occasionally, remains of selachians with fin spines also show this position, probably a result of gases within the body cavity.

Accumulations of fish are often found with remains of terrestrial flora. The carcass assemblage at Smithers Lake is a typical example, as is the arrangement of fish and plant remains in the Upper Devonian at Migasha, on Scaumenac Bay, in Canada, according to Stensiö (156). The fish-rich, oil-producing Middle Eocene Green River Formation of North America also contains remains of an abundant flora: over eighty species of palms, *Ficus,* Lauraceae, Leguminoseae, ferns, oaks, Juglandaceae, *Salix,* and so on.

It is doubtless a consequence of geologic history that the Middle and Upper Cretaceous have so many rich fish localities. Just think of the occurrences in Westphalia, Kansas, Maastricht, Dalmatia and Istria [Yugoslavia], the fish shales of Hakel and Sach el Alma in Lebanon, and the occurrences in Senon and Turon in England, described by Woodward (183): twenty-three selachians, seven holocephalians, fifty teleosts, eighteen ganoids, and two coelacanths. Sandbars are probably often a contributing factor in the mass death of fish because, especially on arid coasts, it is also unstable and dangerous due to sudden changes in salinity, poisoning from sulfuric acid, freshening of water, evaporation and formation of salt deposits, encroachment of the sea, and so on.

In Alaska, salmon have ended up in pools created by the flooding of the Yukon, and their carcasses are buried in layers in the clayey flood sediments. Porsild (131) made us aware of the mass death from physiological causes of the fish called capelin (*Mallotus villusus*), whose carcasses carpet the calm bays of Disko Island [Greenland] after reproduction. This phenomenon explains the frequent occurrence of this fish in the *marlekor* concretions of recent sediments in Norway and Greenland, which have been well documented as far back as Agassiz (10, vol. 5, pl. 60). Similar occurrences are found on the west coast of South America.

Whole schools of small fishes such as *Palaeoniscum freieslebeni* in the Kupferschiefer, *Leptolepis sprattiformis* [*Leptolepides sprattiformis*] in the Upper Jurassic limestone in Bavaria, and small pholidophorids in the asphalt shale of the Upper Dolomite levels at Seefeld, in the northern Tyrol [Austria], are found fossilized. Schools of fishes with the individuals lying parallel to each other are also seen near Sendenhorst [West Germany] and in Lebanon. The concentrations of fishes in the sediments of arid areas are characteristic. We have often found remains of Triassic fishes simply burrowed into the sediments. The position of the *Semionotus*

in the Buntsandstein is almost unvarying. In almost every public collection, there are slabs from Coburg on which the fish are often arranged axially parallel and form a band or swath. There is an interesting slab in the Geological Institute at Göttingen on which four fish skeletons lie completely axially parallel in a row, one above the other. Each fish is one handbreadth away from the next, but the heads point toward different sides, probably due to constant oscillation of the water. Two other specimens from the same collection are also lying axially parallel one above the other, but head to tail and close enough to touch. Exactly the same concentrations in bands or swaths are seen in the well-known slabs of *Semionotus* from South Africa. The slab shown by Abel (4) from the Senckenberg Museum shows twelve individuals axially parallel, most of which are not touching each other.

Hennig too (67) described a slab with eight specimens of *Semionotus capensis* Woodw. Seven are lying parallel, one behind the other, and a single one is lying at an acute angle to the others. Hennig emphasized that the fish do not usually overlap and went into detail on the subject of how certain burial conditions are characteristic of certain kinds of rock. On the other hand, the slab described by Schellwien (139) shows seven individuals deposited in such a way that the heads or tails of several form a new front against which other individuals lay crosswise.

Deecke (38) has more than once discussed the close relationship between the composition of the rock and the fish remains found in them. He maintains that we can expect whole skeletons only as a result of calm water with undisturbed sediments, and even then only if no scavengers were present. He says (Deecke 40) that the fish shales, which I would like to tie to the presence of sandbars, are the bituminous, putrifying mud sediments of quiet lagoons or bays. As examples, he cites the Perledo limestone, the asphalt shale at Besano, Lumezzane, Giffoni, Seefeld, and Adneth, the clay shale at Raibl, the coal shales in New Jersey and Connecticut, and the fish horizons at Spitzbergen.

Johannes Walther (171) emphasizes in *Allgemeine Paläontologie* the interesting biological problem posed by fish localities still today and cites the localities at Oesel, Bundenbach, Mansfeld, Perledo, Raibl, Seefeld, Lyme Regis, Solnhofen, Lesina, Lebanon, Sendenhorst, Monte Bolca, Ulm, and Messel. This list could be extended but includes the most important names.

When carcasses of alligators that have lain on dry land are severely desiccated, they show a pronounced S-shaped or circular curvature. This phenomenon diminishes as the place where the carcass lies dries out after the water recedes. Plate 16, figure C shows the carcass of a coyote with sharply flexed hind legs and completely turned head—the effects of severe desiccation. And when now and then a carcass that has lain in water for a long time comes to rest in a circular or bowed shape, such a typical indication of mummification is of considerable geologic interest. If we look at

the copperplate engravings (of saurians from the Kupferschiefer) made as early as 1856 by H. von Meyer (108), we must acknowledge that none of the twenty-one individuals buried in the sediments is uncurved. To make it even more clear, I give a number of outline drawings of these skeletons (pl. 35, figs. A–E). Often the spine forms a closed loop, the spinal processes facing inward (H. von Meyer, pl. 1, fig. 1, and pl. 7, fig. 7). However, the same is true for a small saurian from the Kupferschiefer in the Göttingen collection, and we see the same position (pl. 17, fig. B) in a find made just this year at the Wolfschacht near Eisleben. Furthermore, additional saurian discoveries from the Kupferschiefer show exactly the same position, especially *"Aphelosaurus"* [*Weigeltisaurus*] completely curled around on itself, which Jaekel will describe. No other position seems to occur in these saurians.

If we compare a deposit such as that of the Holzmaden Posidonia Shale with the saurians from the Kupferschiefer, we often find animals in a completely straight and natural position, with the skin either preserved or in all stages of natural decay, and disarticulation has usually been caused by the carrying-off of bones by scavengers. Highly curved skeletons are unusually rare. It is true that the teleosaur (fig. 22) shows a certain curvature, but this can most easily be explained by the motion of the water and the pull of the ligaments.

Some remains, however, indicate previous desiccation and mummification. In 1908, Pompeckj obtained a small *Pelagosaurus typus* from the upper Lias at Holzmaden for the collection of the Geological Institute at Göttingen. The skeleton lay on the underlying stinkstone. The lower jaw is detached on the outer side, and the torso is strongly curved so that the tail would have touched the head had it not changed direction: at the highest point of the pelvic girdle, the tail bends at almost a right angle, points rearward again as far as the bend in the tail, and then bends again in the direction of the skeleton, so that it points toward the pelvis.

In the museum at Trieste [Italy], I saw a *Mystriosaurus* from Holzmaden whose bones formed a complete circle. The position is even more pronounced in an *Ichthyosaurus quadriscissus* Qu. emend. E. Fraas [*Stenopterygius quadriscissus*], also from the upper Lias from Holzmaden. It was acquired for the paleontological division of the natural history collection at Stuttgart [Staatliches Museum für Naturkunde] by Dr. Berckhemer, to whom I am grateful for permission to publish these interesting remains (pl. 17, fig. A). The creature, on a rectangular slab 38 × 40 cm, is exposed on the ventral side and also forms a complete circle, with the head lying inside and the tail curling above and past it. The head and the first vertebra of the neck are almost dorsal, but then we notice a twist in the long axis of the vertebral column that results in the ribs of the left side of the body being thrust toward the inside. Under pressure from the skull, the ribs in the upper group are less splayed than those separated from them and lying beneath. In the upper group, the ends of the ribs diverge, and in

the lower, they converge. The ribs of the right, outwardly curved, much longer side of the body have separated due to excessive stress and, more or less detached, lie close up against the body. The vertebrae lie obliquely like a roll of coins, often with changing directions. At the points where the direction changes—the middle of the dorsal spine and two places in the caudal spine—they are perpendicular. We are probably correct if we assume that after death and before burial, this carcass lay dried up on a beach and was later lifted off and transported again; we must also assume that this form of burial was typical for the carcasses of *Protorosaurus* in the Kupferschiefer.

This circular skeleton is also characteristic of animals that died in coastal silts, especially in the lower latitudes where they would have been exposed to strong rays of the sun during ebb tides. At the mouth of the Calcasieu River, between the small fishing town of Cameron and the Gulf Coast, I observed in the silts of an oyster bank at the edge of a channel affected by ebb and flow one such circular carcass of a horse, with the ribs in the characteristic position. The circle of bones had already been colonized by young oysters, and the shape of the bone or tooth on which an oyster grew was molded right into its upper valve; as the edge of the shell grew free beyond its point of attachment, it bent down and grew normally.

The skeleton of the *Trachelosaurus fischeri* preserved in the museum at Halle, described by Broili and Fischer (28), is also sharply curved; it decayed considerably after the curvature had formed. This is easy to understand, given the frequent changes of water level in arid regions. The skeleton is strewn over an area between two and three meters square. It lay on clayey mud and as it dried out bent completely around toward the inside, so that the cervical part of the vertebral column lay facing the dorsal side of the rest of it, exactly as in the saurians from the Kupferschiefer; gastralia are strewn over the whole slab. The many clay nodules clustered around indicate that in fact this area had once been a shore. From the part of the spine that is still articulated, Broili could ascertain that it described a tight three-quarter circle, the center of which was covered with many of the ribs of the right side of the body. In the left half, many of the ribs lay on top of the bodies of the vertebrae. Articulated postsacral vertebrae are seen at the beginning of the caudal area. In his explanation, Brioli considers it a floating water carcass. I think of it more as a mummified shore carcass that became disarticulated as the water rose.

In the case of the Kupferschiefer, Pompeckj (130) has already expressed himself clearly, saying that our *Protorosaurus* remains can only be washed-up carcasses; it is easy to understand that, as a consequence of heavy exposure to the sun's rays and perhaps of an abundance of salt on the shore, the carcasses of the *Protorosaurus* were mummified in a sharply curved position and, as the water level rose, were carried from the shore into the basin, where remains accumulated continuously. Ripple marks, fold structures in the sandstone, and, more commonly, desiccation cracks are

found frequently in the subjacent beds of the Kupferschiefer of the Mansfeld syncline. Most important, the bits of plant material lying parallel in the copper-containing Frankenberg Zechstein represent a coastal phenomenon: the mineralized plant remains clearly appear to have been deposited systematically in strandlines. We also have such examples in the Solnhofen lithographic limestone (see the section on curvature of the neck).

At this point, the fish- and reptile-rich beds of the Lower Cretaceous at Lesina, in Dalmatia, should be mentioned. In the museum at Trieste, there is a *Carsosaurus meresetti* from Comen, whose body forms a loop with the neck and tail running along next to each other. Just as clear is the curvature in the *Opetiosaurus bucchichi* (fig. 24) described by Kornhuber (94); here, the left half of the body forms the convex outer side, causing the ribs to point toward the rear at such acute angles that they press against the trunk, while the right side of the body is concave, and the ribs splay at right, or even obtuse, angles from the backbone and diverge toward the inside. The head has become separated from the neck, which is curved around to the left, and the lower jaw is displaced toward the right. The tail, like a chord of the half circle formed by the torso, cuts through the space between the neck and the skull. The creature lies dorsal, but the vertebrae of the neck are turned over. The observer sees the ventral side of the lower spine and the lateral surfaces of the fore- and hind ends of the vertebrae of the neck, which is curved sharply left and rearward. Kornhuber thinks that at the time of burial there was water pressing against one side, which caused the head and the two forward vertebrae of the neck to be wrenched from the rest of the skeleton and to move to the other side of the outstretched tail. He also believes that water pressure caused the ribs on the left side to press against the spine and those on the right side to spread, and the forelegs, including the right scapula, to separate and pull out.

The skeleton of *Procolophon* from the Triassic, described by Seeley (Huene 72; see fig. 23), also shows spinal curvature as a result of desiccation. The torso and the upper part of the tail are gently curved, concave on the left, convex on the right, and both the tail and the neck are bent back toward the right convex side of the body so that they approach each other. There are also some sharply curved skeletons among the Credner (32) material from the Rotliegend of the Plauenscher Grund near Dresden [East Germany]; for an example, see his plate 24, figure 1, and plate 25, figure 1, and further, the skeleton of *Palaeohatteria longicaudata* Credner [*Haptodus baylei*] in his plate 24, figure 1. There is not space enough to mention all the examples of this type of position.

On the skeleton of *Opetiosaurus* (fig. 24,) there is a peculiar displacement of the spine, which we often find in the neck and in the tail and sometimes also in the torso. The torso of the original bloated carcass is much thicker than the neck and the tail. The skull is often flattened and lies with either the under or the upper side flat on the ground. Consequently, at the place where, owing to the shape of the body, stress is greatest, moments of rotation and tension arise, which cause the skeleton to break apart. Figure 27 shows a specimen of *Branchiosaurus amblystomus* Cr. [*Apateon* sp.] from the Rotliegend of Friedrichroda [East Germany] after Langenhahn (97, pl. 1, fig. 1). The tail and torso we see in profile, but the flattened head lies upside down. The same author also had a *Protriton* [*Apateon* sp.] skeleton in his collection (fig. 26), in which the tail, part of the pelvis, and the torso are seen from the dorsal side, but the whole pectoral girdle and the forelegs have turned ninety degrees from the axis of the body and are seen from the front. This twisting also caused the head to turn; furthermore, the head bends upward ninety degrees, causing the skeleton to lie on the bedding plane in an odd position. Kellner's valuable excavations provided the collection with a number of other interesting skeletons showing the same type of curvature. A whole series of such displacement phenomena could be displayed using these specimens.

Currents often bend the neck and the tail in the same direction. After the convex torso has collapsed completely, and often asymmetrically, the neck and tail are often dislocated and bent in the same direction by currents. We see this situation reflected in a *Lariosaurus* Curioni from the Perledo Muschelkalk at Lake Como [Italy], in the Senckenberg Museum in Frankfurt, and shown by Abel (3, fig. 382, p. 484).

We often observe that long tails follow the curvature of the rest of the body up to a certain point, but then bend S-shaped to the other side. This position was not necessarily caused by desiccation. Such a passive, nonextended position is completely plausible as one of relaxation, in which the antagonism of the muscles has ceased—just as the curvature of many fish carcasses corresponds to the relaxed position their bodies assume when they leap out of the water. *Datheosaurus macrurus* Schroeder [*Haptodus baylei*] from the lower Rotliegend near Neurode [now Nowa Ruda, Poland] shows such a curved spine. The axis of the S-curve is almost at right angles to the gently curved axis of the body. We often see this arrangement in animals in purely ventral positions. *Scincosaurus crassus* Fritsch from the Upper Carboniferous gas coal at Nürschan [Nyrany], in Bohemia, shows a similar S-curve of the tail, which lies parallel to the right side of the body.

4 The carcass assemblage at Smithers Lake and its origin

1. THE SIGNIFICANCE OF THE CLIMATE

To understand anastrophe, that occasional natural event that wipes out large portions of the animal world and, depending on its frequency, regulates the propagation of species and displaces biotic communities, we need a more exact knowledge of climatic conditions, which are also significant for many other aspects of geology.

The state of Texas is one and a half times the size of the original German Reich. Because of the state's size, when considering the weather it is useful to divide Texas into sections—west Texas, northwest Texas, central Texas, and east Texas—so that climatic conditions do not become indistinct through overgeneralization. As interesting as the rest of the state may be climatically, in this study we must limit ourselves to the meteorological data of the Gulf Coast.

The area around Smithers Lake belongs climatically to the stretch of land usually known as southeast Texas and includes the coastal region and the Rio Grande valley. Within this area of 60,000 square miles (96,600 square kilometers), there are relatively many weather stations because the population, mostly farmers, is expanding rapidly owing to increasing cultivation of rice using artificial irrigation and the growing of early vegetables. The weather stations closest to Smithers Lake are located at Rosenberg, Sugarland, and Columbia; the figures in table 1, which are based on twenty years of observations, will give an idea of the regional climate.

The prevailing winds in Columbia come from the south or the southeast, with northerly winds predominant only in September and, to a lesser degree, in November. The average annual snowfall for the city of Houston

TABLE 1

	Weather Stations (city and county)		
	Columbia (Brazoria)	Rosenberg (Fort Bend)	Sugarland (Fort Bend)
Elevation			
ft	34.0	108.92	49.0
m	10.36	32.92	14.9
Annual average precipitation			
in	43.81	57.24	42.69
cm	111.0	145.24	108.43
No. days with more than 0.01			
in rain	65	76	85
(.254 cm)			
Temperature			
Annual average			
°F	68.9	—	68.5
°C	20.5	—	20.3
Maximum			
°F	79.2	—	79.9
°C	26.22	—	26.5
Minimum			
°F	58.9	—	57.7
°C	14.94	—	14.28
Highest			
°F	102.0	—	108.0
°C	38.9	—	42.9
Lowest			
°F	5.0	—	8.0
°C	−15.0	—	−13.3

(near the coast) is 0.1 inches (2.25 mm); for Galveston (on the coast), 0.4 inches (10.2 mm). The average annual wind velocity for the Houston area is 8.2 miles per hour (13.196 km); for Galveston, 10.7 miles per hour (17.219 km). The amount of sunshine annually, given as a percentage of the highest possible, is as follows: Houston, 63 percent; Galveston, 64 percent. The average humidity in Houston at 8:00 A.M. is 86 percent; at noon, 57 percent; and at 8:00 P.M., 87 percent. The figures for Galveston for the same times of day are 83 percent, 71 percent, and 78 percent.

The most important data are the frost dates. Records have been kept in Columbia for the past twenty-one years. The average date of the first killing frost in the fall is November 25, and the last in the spring is February 27, leaving a growing season of 271 days. The earliest recorded date of the first killing frost in the fall was on October 27, and the last in the spring on March 24. Houston has been keeping records for nineteen years. The average first and last killing frost dates there are November 29 in the fall and February 16 in the spring, leaving 286 days frost-free. The earliest freeze came on October 25, the last on March 26. Sugarland has records going back twenty years. The average first and last frost dates are November 25 in the fall and February 25 in the spring, leaving 273 days frost-free. The earliest recorded frost was on October 30, and the last on March 29.

On the average, the temperature immediately on the coast seldom rises above 100°F (38°C) in the summer and only infrequently does it fall as low as 10°F (−12°C) in the winter, but further inland temperatures often rise above 100°F (38°C), and in the Rio Grande valley, sometimes above 110°F (43°C). The lowest temperatures range from 0°F (−17.7°C) in the north to about 10°F (−12°C) in the south, but these are exceptional. On the coast, light frost is possible from December through February, but true killing frosts are limited to a few short periods, and in many years there are none. In central Texas, killing frosts may occur between the end of November and the middle of March.

Rainfall in the coastal districts shows variable daily, monthly, seasonal, and yearly amounts. Daily amounts of 16 inches (40 cm), monthly totals of 20–30 inches (50–75 cm), and yearly totals of 60–80 inches (150–200 cm) have been recorded, but there have often been months when not more than a trace has been recorded, and years when scarcely more than 20 inches (50 cm) have been measured at Galveston and less than 6 inches (15 cm) at Corpus Christi. Annual precipitation diminishes going west from the coast and south toward the Rio Grande; at Beaumont, the most easterly city of this area, the annual average is almost 48 inches (122 cm), whereas at Fort Clark, the most westerly city, it is 22 inches (55.9 cm); the northern part of the central district averages roughly 32 inches (81 cm), while in the central Rio Grande valley the average is less than 20 inches (51 cm). The amount of precipitation continues to increase to the east on into Louisiana; from Louisiana west there are interesting distributional boundaries for flora and fauna that are closely tied to the dictates of available moisture. In this connection, I mention not only the occurrence of the epiphytic *Tillandsia* (Spanish moss), generally found along the Gulf Coast from Mississippi into east Texas, but also the riparian woodlands that extend beyond the Colorado River and finally disappear completely farther west. An exact establishment of distributional boundaries would surely contribute significantly to a better understanding of the climate.

West Indian hurricanes can occur near the coast during late summer and early fall, but during the rest of the year here and in areas lying inland, severe storms seldom occur because prevailing winds blow from the south and east. During the winter months, however, severe storms accompanied by north winds and sudden, very pronounced changes in temperature can be expected. This weather pattern is most likely to occur inland, but sometimes these northers reach the coastal region, as one did in the winter of 1925–26, and this one moved as far east as Florida. To evaluate such events, it is essential that one be familiar with the most important elements of weather observation. I can present only the most basic ones here and must refer the reader to the reports issued by the U.S. Department of Agriculture for additional information.

The climate on the Gulf Coast is somewhat like that of the tropics in that most of the rain falls in the autumn and the winter. In Florida, one can even speak of a true tropical rainy season, which contrasts with a summer of such pronounced dryness that artificial watering is necessary, or at least desirable.

In the rest of the Texas lowlands, the frequent summer droughts, which lead to harvest failure, are reminiscent of tropical seasonality. In 1925, there was a drought that extended from San Antonio to Taylor, Texas. At the time I visited Taylor, it had not rained for eleven months, and nothing green was to be seen. Mirages formed everywhere. At such times, the enormous importance of the spring horizon at the boundary between the Taylor Marl and the underlying Austin Chalk becomes very clear. The water in these overflow springs is quite cool. Extensive evidence of Indian activity is found at all these places, not just the beautifully crafted weapons that were left with the dead, which we always see in collections, but also real tools, such as tomahawks, scrapers, drills, and knife blades. The old trails of the Indians followed the strike of the spring horizon, and it is a widespread popular belief that certain crooked trees seen today were bent that way by the Indians to indicate the direction to the next source of water.

During such a drought, the cost of feeding livestock becomes exorbitant, and the farmers, who are poor anyway, cannot make ends meet. To make matters worse, during the most recent drought, some of the water holes became unusable because of their high alkali content. The price of a cow finally dropped to only two or three dollars, and still there were few buyers. A few clever fellows loaded their livestock onto trucks and drove them to the Gulf Coast, where there was still rain, thereby saving their animals from starvation.

It was drought, too, that brought about the colossal mass death of cattle on the Gulf Coast at the end of 1924. A temporal succession was evident: in the winter, there was death in the south from the icy wind, and survival in the north because the animals there were hardier; but in the summer, there was death in the north, owing to the effects of the drought, while on the coast, except for the infections brought about by the masses of carcasses left over from the winter and the other usual diseases, conditions were good and the animals throve. (People did try to get rid of the carcasses—they either buried them or doused them with oil and burned them, at least in the vicinity of the large cities, where it really mattered. But farther out in the country, it looked quite unpleasant, for the carcasses lay around all year long.)

A year later, after the putrefaction and drying-out of summer, one could still see direct evidence of the fatal catastrophe. Enterprising persons became dealers in bones; they trucked the remains to train stations and shipped them out by the carload. Mountains of these bones piled up at the small stations in the Louisiana woods offered a stark reminder of the relentless cycle of life and death. After some time has passed, one can also look at the economic consequences objectively. When there are many cattle, the price is low; when there are only a few, the price goes up. For the most part, the losses even out quickly, for the Texas farmer reckons that on the average 75 percent of his cows will calve.

In the year following the freeze, the indirect consequences were evident. The masses of cattle carcasses attracted scavengers of all kinds. Vultures gathered, as did large timber wolves, prairie wolves, and coyotes. These animals were often shot, and their young carried of as pets.

The reciprocity just described between favorable and unfavorable living conditions in the north and south as it pertains to the cattle, really only marginally domesticated, indicates why the great herds of wild ruminants had to migrate. The migration of the bison can be reconstructed to some extent based on the location of sparsely flowing springs, knowledge of which has been handed down by word of mouth. What really exterminated the bison were man-made, east-west barriers: railways and fences. Every Texas farmer regards barbed wire in the same light as he does the railway, the telegraph, and the Ford car—as a great cultural contribution toward making the south habitable and cultivable; in that vast expanse of land, only barbed wire made it possible to stake claims and establish property rights. So, when the north wind blows and we see the herds of cattle running to the southern fences, only to freeze and die there, we can think of the bison, which met the same fate.

Deckert (37a) cites some further consequences of northers:

The orange was introduced into Florida as early as 1560 by Menendez and has naturalized many times, but the occasional cold waves have been disastrous not only for the harvest, but for the trees themselves. Thus, the culture in question was almost completely destroyed in 1835 and 1894–95, and partially in 1879, 1883–84, 1886, and 1899. Following the freeze of 1894–95, the number of citrus trees fell from 85,000 to 21,000. However, the population always recovers within a few years. In recent times, people have tried to protect against a repetition of such widespread damage by building extensive protective covers for their groves.

Production of raw sugar was first attempted in the Mississippi Delta in 1751; it has been successful since 1793.

Sugar Production in Louisiana

1823	15,000 tons
1834	51,000
1845	143,000
1853	224,000
1895*	355,000

*Best harvest

Of the available arable land in Louisiana in 1899, fully 8 percent (111,000 hectares) was given over to sugarcane. Frosts and other unfavorable climatic conditions have since diminished the yield considerably, so some harvests have yielded only a fifth as much as the better ones. For example, following the hard freeze of February 1899, the yield amounted to only 142,000 tons.

It is understandable that only a slight difference between northern and southern forms of American fauna can be observed. Because there are no mountain ranges to act as barriers, the similarities of climate are reflected in the flora and fauna. Species subjected to sharp contrasts of temperature must be eurythermal. If they are not, they may find themselves unable to adapt even in their usual surroundings. The northern limits of many species are continually advancing and retreating, because although normal living conditions are possible everywhere, every few years, or at even longer intervals, the temperature drops precipitously, and some plants and animals are wiped out. This process is somewhat selective because, for example, deep waters cool down less easily than shallow, food-rich ones, and it is the latter that are so seriously affected by catastrophes. These temperature variations also ensure that the most southerly part of Florida, with its seemingly tropical flora, is not truly tropical. In the winter of 1925–26, a mighty cold wave hit Florida, driving home that fact. The tropical and semitropical fruits cultivated there require artificial protective measures; the American plan to establish rubber plantations there is certainly a questionable one.

Throughout the South the situation is the same. In February 1899, both Charleston and New Orleans had temperatures as low as −14°C, Mobile had −18°C, and Memphis −26°C. The average temperatures for January throughout the area, especially in the interior, were lower than the geographic latitude would imply: Charleston, −8.3°C; Jacksonville, 12.9°C; Tampa, 14.8°C; Mobile, 10.3°C; New Orleans, 12.1°C; Memphis, 4.4°C; Galveston, 11.9°C; and San Antonio, 10.8°C. At Charleston, killing frosts are limited to the period between the beginning of November and the beginning of April.

The bearers of cold waves throughout the lowlands and in the Appalachian Mountains are the northwest winds, which originate in the North American Cordillera and have their most serious effect in Texas. The Siberian *burans* are analogous to the blizzards and northers of North America. When they replace the south wind, they too take a toll of human life and cause unbelievable devastation of livestock.

Winter does not last long in the South, and typically, periods of cold are often separated by longer periods of warmth. However, the degree of cold in often severe. Frost and ice formations are not unheard of, even in the south Florida Everglades. The Indian River and Lake Okeechobee have repeatedly been covered with ice.

An especially dangerous aspect of these storms is the formation of an icy glaze that may happen so quickly that in open country, the telegraph lines snap, not just here and there but between every two posts. Ice accumulates readily on the large masses of epiphytic *Tillandsia* hanging in the trees, and when the branches can no longer bear the load, they bend or break off in great numbers. Cultivated palm trees suffer extensive damage. These storms bring traffic to a standstill because after milder temperatures set in, the roads turn to pure muck. There are human victims too, who usually succumb to an inflammation of the lungs. The following passage is taken from the *Houston Chronicle* of Sunday, December 21, 1924:

Houston crippled by blizzard. Cold wave is due Sunday. Houston under a mantle of ice. Ice covers streets; city is isolated. Train service is demoralized and wire service out in all directions; street cars delayed. In the grip of the lowest temperature in years, Houston was virtually isolated Saturday. Without means of communication with outside points, with a few exceptions, the city was cut off from the outside world. The mercury hovered at 22 degrees, and the cold together with ice and sleet had transportation and communication mediums within the city badly demoralized. By operating cars throughout the night the street railway company was able Saturday to maintain fragmentary schedules on all but a few lines. Telephone and light service was interrupted in many sections of the city.

This contemporary report easily brings to mind the description of one of these storms given by the excellent geographer Theodor Kirchhoff (Deckert 37a):

A few hours before the appearance of a norther, the southwest wind subsides and the air becomes sultry and oppressive. From the north comes a dark cloud, and when it reaches the zenith, the norther breaks. Sometimes it is accompanied by rainstorms, but these do not last long because the cold, dry winds coming from the upper layer of air quickly suck up all the moisture they encounter. At the onset of a norther, both people and animals become very thirsty; their skins dry out quickly and burn and itch. The drop in temperature is considerable and unusually sudden, often going from 24°F to 4°F or −1°F within a few hours; the dryness of the air makes the change even more noticeable. Woe to the unprotected traveler who is surprised by a norther on the open prairie! If he is familiar with the climate of the country, he immediately spurs his horse to a gallop and heads for the nearest house to wait out the storm. With chattering teeth, the inhabitants huddle indoors in front of roaring fires, while outside the winds howl. But as soon as the norther takes its leave, the most magnificent weather often prevails, as if one were suddenly taken from Labrador to Nicaragua. Everyone throws coats and covers aside and goes outdoors to breathe the good air. The fire in the fireplace goes out, and winter is forgotten. For those who do not protect themselves from the storm with warm winter clothing, pneumonia, which claims numerous victims annually, is the punishment. For the livestock that range free during the winter, as is the local custom, these storms are especially destructive. Thousands of head, weakened by lack of food, are unable to resist and succumb to the icy destroyer; in the spring, their bleaching bones lie strewn across the prairie among the fresh, green shoots.

An odd phenomenon occurs when the cold, heavy air of the north wind penetrates the coastal area and pushes the warm gulf air away. The large bodies of water found in the coastal estuaries are relatively warm, and the cold, turbulent air causes waves to form. This produces a genuine fog, especially on the southern shore. See plate 18, figure A, taken near Lake Charles (on whose shores lies the city of the same name) for a typical view of the onset of this phenomenon. This stage does not last long, and soon the southern half of such a lake is shrouded in an impenetrable mist. If the temperature falls below freezing, an icy glaze forms on everything, a disaster for the flora and fauna. If the water is deep enough, the cooling-down may be considerable, but not enough to kill off the things that live in it. In shallow water, however, the mortality of the fauna can assume catastrophic proportions.

Winter along the Gulf Coast is therefore strongly influenced by the sharp opposition of two weather systems—one from the north and one from the south. Because a meteorological divide running east and west, such as the Alps between Germany and Italy, is completely absent, warm air from the gulf may extend as far north as Lake Superior, and cold winter air can penetrate as far south as Texas, Louisiana, and Florida. The sharp contrast between winter cold and summer heat characteristic of the North fades in the area of the Gulf of Mexico, so winters there are usually mild; but the danger of assault by deadly, icy winds is always present.

Let us look at the meteorological conditions for the month of December 1924, when the norther in question loosed its destructive force on the Gulf Coast. First, the temperatures: the highest temperature was 80°F (26°C) on December 18, the coldest 22°F (−6°C) on December 20. Within three days, then, there was a difference of 58 Fahrenheit degrees (32 Celsius degrees). We necessarily expect that the strongest daily fluctuation of temperature lay between these dates, and indeed, a plunge of 45 Fahrenheit degrees (25 Celsius degrees) is seen on December 19. After the initial cooling-off, the lowest daily variation came the next day, December 20, with 5 Fahrenheit degrees (2.7 Celsius degrees).

Precipitation for the month was below normal. There was a little snow on December 19. On December 19, 20, and 21, hail changing to sleet formed an icy glaze on everything and disabled traffic for miles around. The highest barometric reading stood at 30.72 on December 20. Table 2 shows temperatures, relative humidity, and wind velocity.

TABLE 2

		Extreme				
		Maximum		Minimum		Average
Date	°F	°C	°F	°C	°F	°C
12/14	80	26.7	55	12.8	68	19.7
12/15	78	25.6	63	17.2	70	21.4
12/16	80	26.7	67	19.4	74	23.5
12/17	79	26.1	67	19.4	73	22.7
12/18	80	26.7	68	20.0	74	23.9
12/19	68	20.0	23	−5.0	46	7.5
12/20	27	−2.8	22	−5.6	24	−4.2
12/21	31	−0.6	26	−3.3	28	−1.9
12/22	39	3.9	30	−1.1	34	1.4
12/23	51	10.6	36	2.2	44	6.4
12/24	46	7.8	31	−0.5	38	3.6
12/25	37	2.8	27	−2.8	32	0.0
12/26	47	8.3	27	−2.8	34	2.6
12/27	52	11.1	39	3.9	46	7.5

Date	Relative Humidity (%)	Wind Velocity	Date	Relative Humidity (%)	Wind Velocity
12/1	92	21e	12/17	98	18s
12/2	75	33se	12/18	95	19s
12/3	94	19se	12/19	98	22nw
12/4	93	22nw	12/20	94	19ne
12/5	57	16se	12/21	98	16n
12/6	97	20s	12/22	94	8ne
12/7	96	20s	12/23	93	10n
12/8	43	19n	12/24	95	20n
12/9	49	22ne	12/25	47	18nw
12/10	48	21ne	12/26	55	8n
12/11	56	14ne	12/27	52	8ne
12/12	56	9w	12/28	45	12ne
12/13	66	18sw	12/29	96	12e
12/14	87	11w	12/30	97	12n
12/15	78	14sw	12/31	98	7se
12/16	95	16s			

The relative humidity is interesting; the figures I give in table 2 are only for 7:00 P.M. The wind velocity is given in maximum speeds and in miles per hour. The average speed per hour is 8.4 miles and the prevailing wind is from the north. It is typical that on December 15, four days before the norther struck, the wind shifted from the west to the southwest, and then on December 16, 17, and 18, from the southwest to the south. On December 19, the wind came out of the northwest, and on December 20 and 22, out of the northeast; the north wind blew on December 21, 23, and 24; the northeast, again on the 27th; the east, on the 29th; the north, again on the 30th; on the 31st, it finally yielded to the weaker south wind.[26]

2. THE LANDSCAPE AT SMITHERS LAKE

Smithers Lake, which covers about fifteen hundred acres, is the largest of the curious bodies of water found in Fort Bend County west of the Brazos River. Most of the lakes were once river meanders and sometimes maintain a high water level all year because their natural outlets are blocked. On the other hand, they can be completely captured and are then quite full only during large, annual floods. When they are empty, fast-growing, subtropical woodland vegetation takes hold in the nutrient-rich soil and forms an almost impenetrable cover; the fauna is quickly displaced.

The pattern of overflow for the Brazos is unusually erratic. Over and over, the river cuts new beds and bars the old, destroys shores and levees, and floods the land for miles around. There are two main reasons for this activity. First, this river arises in the Cordillera and occasionally carries an enormous volume of water; second, its lower course is so marked by recent uplift that in this area, the river is cutting vigorously. An example can be seen at the rapids between Rosenberg and Richmond, where the calcarenite of the Reynosa Formation causes the river to veer northward until just north of Richmond, where it makes a sharp bend to the south again. This is also the reason that in Fort Bend County the river swings from north-northwest to east-southeast, flowing thirty miles east and only eight miles south before it takes up its former southerly direction again. South of this east-west flowing section, prairie extends right up to the banks of the river, which has cut through here only very recently; the prairie is bounded on the southwest by the Bernard River. At this point, the older courses of the river lie farther north and are indicated by a system of lakes, oxbows, bayous, and a fertile, red, sandy clay bottomland; this bottomland used to be covered with a thick floodplain woodland, but increasingly, the land is being cleared to plant cotton and pecan trees, in part with help from the state prison farms.

Smithers Lake (see fig. 25) lies between Big Creek, site of the oilfield of the same name, which lies atop a salt dome, and Rabb Bayou, which originates right next to the Brazos River near Richmond and joins Big

26. Unfortunately, the important study by Max Hannemann, "Temperatur und Windverhältnisse im Küstengebiet von Texas unter besonderer Berücksichtigung der 'Northers'," did not appear until my book was finished. It is based on observations he made in 1924 and 1925. (*Annalen der Hydrographie und maritimen Meteorologie,* 55th year, vol. 6, Hamburg, 1927.) The author describes, among other things, the same norther as I do in my discussion of the origin of the Smithers Lake carcass assemblage.

Another work has also appeared: W.E. Hurd, *The Northers of the Central American Region: Pilot Chart of the Indian Ocean,* April 1927, Washington.

Creek five miles before the latter empties into the Brazos. Smithers Lake lies in a depression that I take to be an old river meander and is part of the extended area of Smithers Bayou (which empties into Rabb Bayou)—an area exposed to alternating arid and moist conditions, no longer part of the prairie but rather of the floodplain of the Brazos. The east-west course of the river is reflected in the original shape of the lake, which also lay east-west. The shape has been altered, however, because waves driven by winter storms have pounded the southern shore and created a baylike enlargement. This happens when the normal prevailing southerly winds are absent. Consequently, the south and west shores are steeply cut into the "upland flat." This storm-induced enlargement is 1.5 miles [2.4 km] wide and cuts into the south shore 0.5 miles [.8 km]. Therefore, the lake now lies in a pear-shaped depression about 4.5 kilometers long and 2.3 kilometers wide. Although these storms occur only during a brief period every year, their force is enough to ensure ongoing destruction of the southern shores of most lakes in the affected area. The result is that the lakes are continously migrating southward, and as they do so, even the strongest virgin woodland is washed out from underneath and cannot prevent the displacement of these large floodplain lakes. The same thing happens to small, temporary, prairie lakes: the southern shore is always steep and the northern ends are finger-shaped and very shallow. This migration can lead to a curious capture of rivers. The sluggish Bayou Teche near Jeanerette, Louisiana, is threatened with capture by Grand Lake.

The carcass assemblage that I am about to describe is indicated on the map in figure 25 by a dark shaded area; it is shaped like a half moon, coming more gradually to a point to the west, and lies in the most southerly part of the bay on the south shore. It is about 1,300 meters long by about 350 meters wide at the widest; its area of about 200,000 square meters makes it a vertebrate locality of considerable size.

To the east and even more to the west, where the border of stumps becomes wider and wider, lay many empty shells of the huge *Anodonta,* often still paired. They were found much less frequently within the carcass assemblage. The whole surface was covered with huge garlands of intricately assembled pieces of driftwood and carcasses, with bays and recesses caused by the angular positions of the various elements, between which were smoother surfaces upon which a few things were strewn here and there; the whole reflected many different water levels with changing, but continuously abating, currents. The originally loose debris had been washed ashore in such a way that its individual elements propped each other up and reinforced each other, thereby offering resistance to the force of the waves. In January and February, a large part of the area that was exposed later, in March, was still under water, although the level was going down. Eleven months later, on November 22, 1925, I could scarcely believe my eyes when I saw that almost the whole lake bed was covered with vegetation five meters high. The huge carcass assemblage was hidden by a thick stand of "coffee beans" [mesquite], a fast-growing, ubiquitous member of the Papilionaceae, which grows twice as tall as a man, or more. Some of the leaves had already dropped, and the ripe seeds in their dry pods rattled in the wind. These dry seeds are always ready to be borne ashore at the next high water level, where they germinate in the exceptionally fertile soil, forming borders of seedlings. The swaths of driftwood and carcasses were covered thickly with grass, and only here and there did a bleached turtle shell or fish bone protrude. It is striking how the slight fluctuation of water level has had so much influence on the appearance of the landscape, filling in the low areas with sediment in such a short period of time.

How many carcass assemblages might there have been before storms eroded the southern shore to its present size and shape? And what has become of them? Most have disappeared, but the floor of the lake must be rich in bones. Out of thousands of carcass assemblages, perhaps only one has been fossilized, but it does happen occasionally, and if the assemblage is interpreted correctly, the information will help decipher the history of the earth. If we study these valuable legacies thoroughly, we will be able to formulate a more exact chronology and reconstruct specific paleogeographic events, as de Geer is already doing in the Scandinavian banded clays: he has reconstructed and dated specific storms. Abel, too, has shown us many new points of departure and ways of studying these phenomena. I hope we are on the threshold of a better understanding of such things. A valuable resource lies scarcely touched at the feet of the researcher.

The western shore of Smithers Lake (pl. 18, figs. C and D) presents a wide surface covered with the rotted stumps of the subtropical woodland that once grew there; among them are countless twigs and branches lying in a more or less orderly way. In the vicinity of the shore, the litter is usually arranged tangential to the shoreline. About twelve years ago, the lake, fairly low at the time, was artificially dammed to raise the water level about three-quarters of a meter, to provide water storage for rice cultivation. The dam turned a lake that had been intermittent by nature into a perennial one. Large parts of the depression extending right up to the remains of an old meander of the Brazos and the stream opening into it were covered with virgin subtropical woodland of the type found in floodplains where the soil is flooded during periods of high water. If such water stands too long, the trees die because the roots are flooded.

On the eastern shore, between this low-lying woodland and the edge of the prairie above it, there is a difference in elevation of about twenty feet. The lack of a wooded area on this shore (pl. 18, figs. C and D) and other indications show that before the lake drained, the natural water level was substantially higher. At the same time, the field of stumps that rotted when the water backed up, seen in the pictures, shows that only a loop in the deepest part of the middle of the lake was free of this type of vegetation, and I think this loop represents a former bed of the Brazos.

At the sloping edge of the prairie, there are live oaks hung with Spanish moss; then there is a zone that is often dry and filled with withered shrubs, which have grown up there where their seeds sprouted. This recent stump horizon formed only within the last ten years and reminds one vividly of the stump horizon of the Miocene lignite in Lusatia [Poland] (pl. 19, figs. A–D, and pl. 26, figs. A–C). In the interior of the lake, trunk and twig remains are lying perpendicular to the shore. The recent stump horizon indicates that the lake was formed only by the displacement of the river and flood damming and not by a change in the climate. In the deeper parts of the lake, sapropel has already begun to cover the stumps. The *Taxodium* of the Miocene died off when its habitat was flooded too long. The same genus living today in North American cypress swamps, with its aerial roots forming fields of arches, is a typical plant of flooded lowlands. This woodland destroyed by rising water offers an excellent example of the completely natural origin of stump horizons and of the lignitic parts of brown coal. So, too, in the central German lignite, the remains of flooded subtropical woodland flora come together with the remains of sclerophyll evergreen vegetation from the "upland flats," an area not exposed to the surrounding flooding.

The swift changes in the course of the Brazos are recognizable everywhere; freshly cut valley walls with collapsing, undercut banks and quickly shifting slip faces successively taken over by bands of vegetation, mostly cottonwoods, emerging from high-water strandlines. The profiles in the valley walls, which often cut across old riverbeds, are reminiscent of the profile of the German Tertiary. They reveal lenses of sand and deposits made in a quieter environment, above which there is often a very black, rich loam buried under reddish flood deposits; the loam corresponds to flood depressions once filled with subtropical woodland. This makes an excellent soil for cotton, if the plow breaks through the reddish sandy clay cover to reach the black layer. Oxbows left behind (pl. 37, fig. C) can dry up and be covered with a stand of trees, but if they are cut off by the accreting banks of the river in its new course, the water rises quite high during floods until only the tops of the trees can be seen above the surface. From the town of Richmond, on the east bank of the Brazos, we can observe this phenomenon in the curved, fish-rich lakes of the vicinity.

The formation of stump horizons in this river environment is entirely plausible. The lake lies in a depression abundantly provided with nutritive material, whereas the higher-lying ridges and watersheds are subject to leaching and are consequently bleached just like some Pliocene weathered soils and certain Spanish soils. There is almost no woodland to the west. This impoverished area is covered with a monotonous pinewoods where the wasteful practice of distilling turpentine is carried out. There is a remarkable parallel between the central German Eocene lignite and the conditions in contemporary Louisiana: Weymouth's and other long-needled pines have often been found with remains of the subtropical, swampy woods. In the Eocene, too, beyond the reach of river activity, we must expect to find sandy, leached soils, poor in nutritive salts, which dried out in summer and supported large stands of turpentine pines; these trees contributed greatly to the increased amounts of bitumen in the coal. I have seen pine needles from the central German Eocene lignite on which one could see with the naked eye all the little furrows filled with yellow-gold resin.

In 1899, 1900, and 1901, the Brazos flooded with such ferocity that people who had lived on its banks (those who managed to escape with their lives) fled and looked elsewhere for a homestead. (During severe floods, the high waters of the Brazos, the Bernard, and the Colorado all come together in the coastal lowlands. The salt dome of Damon Mound rises like an island at the edge of the flooded area, making its suitability as an old Indian settlement very plausible.) Crossing the flooded woodlands of the Bernard River, which appear so harmless when they are dry but which are especially unpleasant during high water, I reached the west bank and came upon the remains of farms that had been left behind. One of them had belonged to seven brothers and had been quite a success at the outset. The men did not suspect that they were located in the high-water area of the Bernard River, which usually ran in a deep channel. The floodwaters surrounded them suddenly; one man drowned and it was only with great difficulty that the others managed to flee, but they had to leave everything behind. They went back north, where they had once sold good land to settle in the south. The disastrous effects of malaria, a disease arising in the swampy area where the floodplain meets the prairie, also caused many families to make the return migration.

At the bar at the mouth of the Brazos, where the water was a good two meters deep, the river was laboriously dredged. A long stone jetty built on the west side of the mouth permitted the water to flow farther out into the Gulf of Mexico, and the natural barrier was strongly influenced by it. Coastal erosion toward the west has already formed a spit from the point of the jetty to the coast, and the dead space thus formed has begun to fill with sediment.

The headwaters of the Colorado River are also in the Cordillera; this river, too, can be wild and rambunctious. Even now, as I myself have seen, when it is full to the banks it carries a stripe of tree trunks down its middle, which dramatically increases its erosive powers and thus its ability to cut new streambeds. In April 1900 it cut through the dam built near Austin in 1892. Today, more than ever before, people on the Texas Gulf Coast are storing winter water to supply summer irrigation needs. Much hard work remains to be done on this project.

3. LAWS GOVERNING FORMATION OF STRANDLINES

First, a few words on the effect of flow shadows. I have already said that any obstacle—a carcass lying on a bedding plane, for example—divides the volume of oncoming water so that it is compressed on both sides. This compression of the profile leads to a considerable increase in turbulence, so that on both sides, deep furrows can be formed, which promote the sinking-in of the carcass; on the other side of the carcass, the area within the flow shadow is more or less protected from the effects of waves, and sand, seaweed, or other floating material can accumulate there. On the Gulf of Mexico, I once saw a long rib from a palm frond washed up parallel to the shore. As the waves met the obstacle, they divided, were evenly compressed, and flowed around it; a round, subtropical fruit the size of an apple, which had been floating by, had come to rest exactly in the center of the back side of the palm rib. In the museum in the old Akademie in Munich there is a similar fossil example. It is a skeleton of *Homoeosaurus* with extended limbs. The tip of the tail is bent slightly to the right. Between the right foreleg and the right hind leg, just about exactly in the middle of the body, lies a stick shaped like a two-pronged fork; the "tines" point away from the carcass and the other end lies right next to the skeleton without touching it. The side twig of the fork goes off at an acute angle approximately in the same direction as the hind foot. The axes of the reptile and the stick form a right angle. Interestingly enough, the tip of the tail of *Homoeosaurus* is bent in the same direction as the stick, which lies in the flow shadow of the larger remains. Broili interprets this find differently: "Fearing to drown, this terrestrial reptile grabbed the forked stick and floated around on it until it finally succumbed to exhaustion, and drowned anyway. Stick and reptile were washed ashore together and buried together." The drifting ashore together of two objects is often seen because of the attraction swimming bodies exercise on one another. Such a relationship between this stick and this reptile is possible, but it is not necessarily the only explanation; in any event, it is hard to prove. Bits of wood are often found deposited in flow shadows, as the carcass of a large alligator gar with debris accumulated behind it shows (pl. 20, fig. F).

As drifting carcasses and vegetable matter, whose orientation with regard to the shoreline is radial (perpendicular), near the shore, they are pushed around by lapping water so that they come to lie tangential, or even completely parallel, to the shore; the result is a compact shore deposit of great resistance composed of originally separate elements. It is by no means haphazard in its construction. If the object washed ashore is too bulky or irregular in shape, one of the splayed parts anchors and the rest swings around this fulcrum in the direction of the shore. Often, only the center of gravity serves as a fulcrum. If the pieces are so heavy or bulky at one end that they anchor too securely, they remain perpendicular to the shore. Thus, the largest tree trunks of the driftwood zone of the Gulf of Mexico do not lie parallel, or even tangential, to the shore, but perpendicular to it. The roots are always distal, that is, directed toward the sea. It is the same with huge trees on the southern shores of the large lakes in the floodplain of the Mississippi in Louisiana.

Let us look again at the strandlines of the carcass assemblage at

Smithers Lake. They are composed of vertebrate carcasses and pieces of wood arranged in an orderly fashion. In the foreground of plate 26, figure A, there is the carcass of a large ganoid fish lying tangentially to the shoreline, and parallel to the fish and in an extension of the axis of its body lies a system of pieces of wood. The head of the fish and the ends of the sticks form a new front, which now itself serves as a shore, so that another system of wood and fish carcasses attaches itself to it almost at right angles, and with it is the carcass of a large alligator gar, whose axis is perpendicular to that of the ganoid fish. Then there are more sticks going in the former direction, and so on. The same phenomenon is seen on certain *Semionotus* slabs, on the *Neusticosaurus* skeletons (pl. 32) from Eggolsheim [West Germany] and in the fine group of *Pantelosaurus saxonicus* [*Haptodus baylei*] from the Rotliegend (see fig. 28). The first example is relatively easy to overlook, but the larger strandlines and garlands are put together in this same way. The interference of waves coming from different directions is also an important factor. These strandlines, therefore (see pl. 25, fig. C, and pl. 26, fig. B), are not only very orderly, but are truly in harmony with the force that drove them ashore. The same sort of balance is seen in a shell pavement or on a sandy beach that is covered with heavy stones at places where there is turbulence. Originally, the components are loose, unconnected, and easily moved about. But the force of the waves must be met with resistance greater than itself if a shore deposit is to form. So the material is pushed together and piled up in an orderly way, just as in strandlines of organisms I have described (172a) on the North Sea. Furthermore, the whole process of material drifting ashore and forming strandlines and garlands is nothing but the advancing of the shore at the expense of the water. Plate 26, figure B shows one such strandline in which a triangular arrangement developed from material that originally lay almost completely parallel to the shore, mainly through the addition of shorter pieces in the middle of longer components.

Plate 29, figure C shows us a typical twig pavement with pieces added in a firm, axially parallel position. It is interesting, by the way, how similar this situation is to fossil occurrences. Stensiö (156), from Stockholm, explained to me that the locality he visited for Upper Devonian fish from Migasha, at Scaumenac Bay [Quebec], in Canada, showed the same arrangement and regularity. The comparison with *Semionotus* slabs can also be made. On them we observe that chevrons and changes of direction are found, as seen in plate 25, figure D, but the direction and position of these seemingly atypically placed remains exactly fit the spaces between the fish carcasses.

In considering how such objects are washed ashore, we must note that even when the wind is blowing at an acute angle, the shore waves almost always hit the beach head on; only seldom do we see a different situation, one that can easily lead to coastal erosion and formation of spits. The bottom of shallow bodies of water has a braking effect, so the closer the waves come to the shore, the more they turn to face it. If one follows the courses of waves in a body of water, one recognizes a zone where they make a curious swing, the crests turning until they line up parallel to the shore. If the waves are large, this shift takes place at a considerable distance from shore. If the waves are coming ashore at an angle, the side that hits the shallow water first is immediately slowed down, while the other side continues to move freely.

4. GANOID FISHES AT SMITHERS LAKE

By studying the gars, or garpikes, at Smithers Lake, we gain a much better understanding of how the older ganoid fishes, which occur often as fossils in Europe as well as in America, were preserved. Even the Egyptian Cretaceous contains their remains. They have been considered a highly specialized relic of *Semionotus*-like fishes. The most well-known occurrence of garpike in Germany is in the middle Eocene bituminous shale at Messel, near Darmstadt. Isolated scales and larger pieces of the scaly hide have been known in this region for a long time. More recently, whole skeletons of this "North American" ganoid genus *Lepisosteus* have been found. Some time after the first find at Messel, the scales of this fish were also found in the Late Tertiary *Corbicula* marl of the Schleusenkammer, between Frankfurt and Niederrad, and were described by Kinkelin as *Lepisosteus straussii* [*Atractosteus strausi*]. This species was almost twice as big as the garpike *Lepisosteus osseus* of the waters of temperate North America—considerably smaller, therefore, than the alligator gar *Lepisosteus ferox-viridis* [name no longer valid]. Garpike remains were also found in the lower Miocene at Böhmen. The fact that scales of these gars have also been found in the coarse Eocene limestone at Paris need not imply that the habitat of living gars is different; even today they inhabit not only the rivers but all the brackish lagoons along the Gulf Coast and are regularly found in the "cuts," those places where the sea is directly connected to the lagoon. From there, gars may happen into shallow coastal waters, and I have often found their carcasses on the coast of the open gulf (see pl. 14, fig. A). The occurrences in the Oligocene at Osterwedding and Westeregeln [East Germany] probably came from the same environment. With the exception of those at Messel, the seven or eight species of garpike occurring in England, France, and Germany from the lower Eocene to the Miocene are based on incomplete remains, mostly scales, vertebrae, and skull fragments. Thus, it was all the more important when Eastman (44) described and drew two complete fossil skeletons from America. To be sure, O.C. Marsh (106) had already described two fossil species of gar from the Wyoming Eocene: *Lepisosteus glaber* and *Lepisosteus whitneyi* [names no longer valid]. As to the single identifying characteristic, however, he said of the one species only that it had unusually short vertebrae, and of the other that the vertebrae were relatively longer. In the meantime, Leidy and Cope (100) also described a number of species based on more or less incomplete remains—isolated vertebrae, scales, and parts of the skull. More complete was the *Lepisosteus cuneatus* Cope from the Miocene in central Utah. By 1900, four species were known from the American Eocene and two from the Miocene. Of these, I cite the following:

> *L. atrox* (= *L. anax* Cope) [*Atractosteus atrox* Leidy]; Middle Eocene, Wyoming
>
> *L. simplex* Leidy [*Atractosteus simplex* Leidy]; Middle Eocene, Wyoming
>
> *L. notabilis* Leidy [species no longer valid]; Eocene, Wyoming
>
> *L. (Clastes) cycliferus* (Cope) [species no longer valid]; Miocene, Central Utah
>
> *L. (Pneumatosteus) nahunticus* (Cope) [species no longer valid]; Miocene, North Carolina

One peculiarity of the two complete Eocene skeletons described by Eastman is that they have no primitive characteristics. Instead, there is a striking, unmistakable similarity to species living today; Eastman thought they had appeared very suddenly at the beginning of the Tertiary, completely specialized and without any transitional feature that might give a clue as to their origin. In this regard, I should point out that as early as 1887–88, Reis (133) had already done interesting studies on *Belonostomus* and *Aspidorhynchus* and their relationship to living *Lepisosteus*. He seems to have discovered at least some points of relationship based on the following observations: In *Aspidorhynchus* the cross-jointing of the rays of the unpaired fins is not as sharply defined, but the spine provides interesting grounds for relating *Belonostomus* to *Lepisosteus*. In *B. tenuirostus* Agassiz the high vertebrae already have the typical hourglass shape and are significantly longer than high, just as are the very large vertebrae in the tail of *Lepisosteus*. The tail fin of *Lepisosteus* shows the very ancient development of internal and external heterocercality, whereas the two fossil species show internal heterocercality, but externally the fin is divided into two equal lobes. According to Reis, then, the *Aspidorhynchus* must be very closely related to *Lepisosteus*. There is almost no difference between the skull bones of *Aspidorhynchus* and those of *Lepisosteus;* even the small nasal bones, which run from the head above the operculum to the clavicle, are the same. The sculpture of the headbones is also analogous. The structure of the *Lepisosteus* upper jaw is not found in the others. In *Lepisosteus* the large teeth bore through the upper and lower jaws; this is seen especially in the upper jaw behind the nasal opening, where there are two or three holes through which the lower-jaw teeth penetrate to the dorsal side.

Assmann's conclusions (18) are the exact opposite of Reis's; rather, he agrees with Eastman (44). I should like to quote him, too:

The rostrum of *Lepisosteus* is formed on the outer side by the much extended frontals, the nasals, the very small premaxillary, and the maxillary, and on the inside by the unusually wide, forward-extending parasphenoid, the small vomer, the palatine, as well as by the ecto- and entopterygoids. The rostrum from *Aspidorhynchus* shows an entirely different structure. It is formed on the outside by the mesoethmoid, the nasals, and the premaxillary, on the inside by the lateral ethmoids and the vomer. Furthermore, the roof of the skull of each form is quite different—in *Lepisosteus* it is very wide, in *Aspidorhynchus* very narrow. In *Lepisosteus,* the squamosal, lacking in the *Aspidorhynchus,* lies to the side of the parietal. Whether it is absolutely not present in *Aspidorhynchus,* or whether it is just that the parietal has been shoved over on top of it could not be determined.

According to Assmann, the main similarities between *Aspidohynchus* and *Lepisosteus* occur in their outer bodily form and the outer sculpture of the bones of the roof of the skull. Further, there are certain similarities in the position of the fins, the pectoral girdle, the hyoid arch, and the gill arch. All these similarities are, however, probably attributable to convergence. Nevertheless, *Aspidorhynchus* will have to be left in the Rhynchodontidae [name no longer valid], even though a close relationship to living gars does not seem to have been shown.

Let us look again at the specimen of *Lepisosteus atrox* Leidy described by Eastman (44). It is a large species, similar in size and general characteristics to Recent alligator gars. The head accounts for roughly one-fourth of the total length; the snout is short and wide. The preservation is typical: two-thirds of the fish is lying ventral, but the abdominal region is twisted around so that between the tail and a point between the anal and abdominal fins there is a clear-cut lateral view. The scales anterior to the line of twisting are in some disorder, but posterior to that point, less so. All the fins except the pectoral are well preserved. The head seems to have been compressed before the skeleton was fossilized because most of the skull bones are displaced; only the jaws and opercular apparatus on the right side have remained intact.

Among the different living and fossil species, the shape of the fin varies.

As a consequence of the bending of the body and the associated disturbance in the anterior part of the scale system, it is difficult to count the longitudinal scale rows or even the cross rows. A cylindrical coprolite 13.5 centimeters long and 5.5 centimeters in diameter was found with the *L. atrox* Leidy. The outside of it is smooth except for some irregular spiral folds. No other particles are recognizable on it, and the whole thing gives "the impression of a healthy digestion," although it is more likely to be the remains of coarser material excreted through the mouth.

The similarity of the skull of *Lepisosteus atrox* Leidy to that of recent

TABLE 3

Species	Number of Fin Rays (radial formula)				Scales in Lateral Line
	Dorsalis	Caudalis	Analis	Pectoralis	
L. atrox Leidy	−8	12	8	6	50–60
L. simplex Leidy	−7	12	7+	—	ca. 45
L. tristoechus (Bl. and Sch.)	7–8	12	8	6	60
L. tropicus Gill [Atractosteus tropicus Gill]	8	12	8	6	52–54
L. platystomus Raf.	8	12	8	6	56
L. osseus Linn.	8	12	7–9	6	62

alligator gars is striking. The fish is lying ventral and the body has been pressed flat; the head, however, has suffered little from pressure and can be studied nearly as well as that of a fresh gar. The shape of the skull is somewhat between that of *Lepisosteus osseus* and *Lepisosteus tristoechus* [*Atractosteus tristoechus*]. The back of the skull of the Eocene form is relatively wider than that of recent forms. There is no noticeable difference between the dentition of *Lepisosteus tristoechus* and *Lepisosteus atrox* except that the teeth in the fossil form are somewhat smaller. Another species from the Eocene of America is *Lepisosteus simplex* [*Atractosteus simplex*]. It, too, comes from the typical Green River Formation in Wyoming. This species is roughly sixty-five centimeters in length, one-fourth of which is taken up by the head. The outer bones are especially heavy and are arranged as in recent alligator gars, but with finer and more granular ornamentation. The ganoid tubercles of the operculum and suboperculum form more or less connecting lines. In *Lepisosteus atrox* and *Lepisosteus tropicus*, the head also accounts for one-fourth of the total length, in *Lepisosteus platystomus* [*Lepisosteus platostomus*], and *Lepisosteus tristoechus*, two-sevenths, and in *Lepisosteus osseus*, with its specialized snout, one-third of the length. Here, too, the body is twisted and the scale system was damaged in the process. The head seems to be almost separated from the body and is completely twisted around so that the ventral side is seen. This position would not have been possible had the viscera not been damaged and displaced.

From the description of both garpikes from the Eocene of North America, it seems that a purely lateral position is not seen. The head is dorsoventral, the tail is lateral, and the whole scaley sheath shows a corresponding, characteristic twisting that is anything but accidental and whose origin will be immediately made clear using recent examples (see pl. 20, fig. D).

There is only one genus of *Lepisosteus* in southern North America, and it is found as far south as Mexico and Cuba. The skeleton is ossified, the tail is heterocercal, and gill-flap rays are present. The fins are reinforced by fulcral scales. The air bladders have parietal cells; the centra of the vertebrae are opisthocoelous; the spiral valve of the intestine is rudimentary. There are auxiliary gills (pseudobrachs) on the gill cover. There is only a single dorsal fin, set so far back that it sits right over the anal fin and gives the creature a pikelike appearance, reflected in the name garpike. The cone-shaped teeth have special widened, heart-shaped tips, which narrow laterally from front to back, and its sizable set of teeth indicates that the fish is an arrant predator. Between the strong fangs there are numerous smaller teeth. The upper jaw is much elongated to form the snout; the nasal openings lie right at the tip. The name "alligator gar" is due to the reptilian appearance of these fishes, especially to the long, bony snout. The creatures supposedly reach a length of twenty feet, but this may be an exaggeration.

There are four species alive today. Of these, *Lepisosteus tropicus* Lyell lives exclusively in Central America. The following three species still live in North America:

1. *Lepisosteus osseus* Leidy. The ordinary garpike or needle gar, as they generally call it on the Gulf Coast, grows to 130 centimeters in length, is olive green, silver underneath, and speckled with black toward the rear. Very young fish show a black lateral band. In the lateral line, there are approximately sixty-two scales. Its snout is twice as long as the rest of the head. Its range extends from the Great Lakes and the Mississippi valley to the Rio Grande, and on the Atlantic and Gulf coasts it can be found from Vermont south. It is usually found only in rivers and lakes, but also sometimes at sea. It is common.

2. *Lepisosteus platostomus* Rafinesque. The fish is usually called the short-nosed gar. It grows to 90 centimeters in length. It is olive green, but darker than *Lepisosteus osseus* L. It has fifty-six scales in the lateral line. The snout is about the same length as the rest of the head, and there is a single row of large teeth in the upper jaw. It is found in the large lakes in the Mississippi valley, and as far south as the Rio Grande. It is more common in warm areas.

3. *Lepisosteus [Atractosteus] tristoechus* Bloch and Schneider. This fish grows to 600 centimeters in length, is greenish in color and paler underneath; there are sixty scales in the lateral line. The snout is about the same length as the rest of the head and it has two rows of large teeth in the upper jaw. It is found in the southern United States and reaches as far north as the Ohio and Illinois rivers.

These three species differ from each other in the following ways: (1) snout twice as long as the rest of the head (*Lepisosteus osseus*); and (2) snout about the same length as the rest of the head: (*a*) with a single row of large teeth in the upper jaw (*Lepisosteus platostomus*); and (*b*) with two rows of large teeth in the upper jaw (*Lepisosteus tristoechus*).

Gars can live in fairly bad water. Air bladders function like lungs, and we often see the fish at the surface of the water gulping air, betraying its presence by a snapping sound. In the spring it seeks out shallow water for spawning. The numerous eggs stick fast to the bottom, and after hatching, the larvae attach themselves firmly to plants or stones by means of a forehead sucker set with many papillae and live off their large yoke sacks for fourteen days. Like *Amia,* these fishes care for their young and are therefore still very numerous despite all persecution.

Gars are the most malicious predators imaginable; they go about their work of killing and eating other fish with a certain smug insidiousness. Once, while standing on a bridge over a coastal river with heavily wooded banks, I saw a long school of small fish swimming upstream. A large gar accompanied the procession and from time to time lunged, snapping, right into the middle of the school, devouring some and leaving a number of others injured. The attack divided the school in two, but even though the gar was still there, the fish kept trying to close ranks.

Fishermen hate the gar with a passion. Many of them, in addition to using the light hand-held tackle appropriate for food fish, set out a strong, heavy line with big hooks, and when they catch one of these predators, they put an end to his menacing behavior with a revolver shot to the head (see pl. 20, figs. A and B). They throw the carcasses back into the water where they are eaten by their surviving comrades. In spite of this effort to get rid of them—Negroes even eat these otherwise despised fish and have found a way to pull off the adhesive scaly hide—reproduction keeps up with the losses, and extermination of this animal is at the moment unthinkable. They seem to be quite long-lived, and when water dries up during drought, they feed on whatever else gathers in the shallow pools, clinging to life with the greatest tenacity after everything else has died.

Since we are talking about the living forms of the genus *Lepisosteus,* we may as well take a look at the other living forms of ganoid fishes. The African *Polypterus* belongs to the crossopterygians, but *Chondrosteus* is a ganoid that is bare and scaleless or has long bony plates on the skin. The ganoids in question, however, belong to the Holostei with ganoid or cycloid scales. Lepisosteids have true ganoid scales. In the family Lepisosteidae, the upper jaw if formed mainly by the premaxillary. The operculum has an accessory plate. There are three branchiostegals and no spiracles. The spiral valve is still rudimentary. There are eight rays in both the dorsal and the anal fins. The Amiidae have cycloid, imbricate scales, a subconical head, and 10–12 branchiostegals. The dorsal fins are long. The tail is convex at the rear and somewhat heterocercal. The single species living today, which is also found in the carcass assemblage I describe, is *Amia calva* Leidy. The Americans call it dogfish or bowfin. It grows to 45–60 centimeters in length, lays its eggs in pits of gravel and sand, and cares for its young.

The recent occurrence of *Lepisosteus* and *Amia* on the Gulf Coast, like the many botanical parallels (I think of recent *Smilax* species, whose tendrils make walking difficult everywhere in the subtropical woodland and whose thorns tear at the skin), ties the German Middle Eocene very closely to contemporary American conditions. And we can thoroughly understand why Andreae had the same impression (8) when he discovered near Messel the massive occurrence of alligator remains with *Lepisosteus* and *Amia*. The area where the Messel oil shale was formed was once a fairly extensive lake and, having been part of the floodplain of a large river system, must have had much in common with today's coastal swamps.

Let us look at the shape of the gar's body and how it affects its appearance after deposition. Plate 20, figures A and B show an alligator gar about 160 centimeters long in a lateral position, photographed from the back; it was caught in Tres Palacios Bay with a heavy line and a big iron hook. The damage to the side of the head was caused by a revolver shot. It is striking that the relationship of its height in the lateral position to its width in the ventral position is 15:23. The contrast is not so great in animals that are still alive, because the body cavity inside its scaley skin flattens out immediately when the fish is laid out, even more in females than in males. The female shown here had about one and a half buckets of roe inside her. Looking at the carcass from above, we notice that the head and torso come to rest naturally in the ventral position, but that the tail fin has a tendency to lie to one side; consequently, the dorsal fins are askew.

Completely natural too is the definite curvature often seen in dead fish, as it appears in plate 23, figure D. Over and over, carcasses found lying in shallow water are curved convexly toward the direction the water is coming from, an indication that moving water pushed the head and tail sideways.

In the lateral position, the torso is the part most resistant to movement, the head and tail hang down, and the skin on the upper side is stretched. Plate 20, figure C shows a photograph I took west of Sargent near Mitchell's Cut on May 17, 1925. The lateral position of this gar is characteristic: the tail and a large part of the torso, including the dorsal and anal fins, are in a lateral position, but the head and snout are clearly twisted and lie dorsal. This is strongly reminiscent of the twisting of the skeleton described by Eastman (see pl. 20, fig. D).

Two carcasses of alligator gar, each 180 centimeters long (pl. 19, fig. D), also showed this natural curvature shortly after they died during the norther. Their concave sides face each other, and we see clearly that the tail end of the closest one is bent around to the lateral position. The water level has gone down considerably from its high point on the day of the storm. The snouts of both fish point toward the middle of the lake, and their tails point shoreward. They were too heavy to have been carried ashore and consequently were not turned around into the shore-parallel position.

If we look at carcasses in a similar position on March 15, 1925, we see that the soft parts have completely decomposed (for example, pl. 24, fig. A). We notice immediately, in addition to the muddy surface, that there is no plant debris; this means that the carcass was originally deposited in water and not on shore. The ganoid scales have held together so well that the tough, scaly skin lies there like a collapsed sheath (see also pl. 20, figs. D and E). The abdomen was originally quite extended because the pressure of putrefying gases strained the scale rows; the effects of this pressure are still visible. Because the head of the carcass (fig. D) pulls the ventral side underneath, the tail assumes a lateral position, causing a characteristic fold from the middle of the neck down the dorsal side of the profile, as is often observed in fossil ganoid fishes. In figure E, the head is also in the lateral position, but there is a typical fold that runs along the back between the head and the unsupported abdominal area. Displacement of the lower jaw, partially due in this instance to feeding vultures, is typical.

Let us compare carcasses of alligator gars that have been washed up to the shoreline. The carcass in plate 20, figure F is in the ventral position; the left half of the body seems the wider of the two halves because the current has caused a pull to the right. Further evidence of the current is seen in the distribution of plant debris and sprouting seeds in the flow shadow of this carcass. The scaly hide has been damaged in several places by vultures. There is a dorsal fold, corresponding to the somewhat asymmetrical position of the skin; the fold becomes more pronounced as it goes toward the back fins, which are reinforced by fulcral scales. It is often found in the ganoids of the Kupferschiefer.

The alligator gar carcass of plate 21, figure A has also been stranded in the water, perpendicular to the shore, in this instance with the head pointing toward the shore. Its snout touches a strandline of plant debris. The left lower jaw, recognizable from its pointed teeth, has been moved off a bit; it lies parallel to the strandline and almost perpendicular to the right lower jaw. We can see by the fins that here, too, the skin lies asymmetrical. The left pectoral fin has been pulled out by vultures. The size of the bits of wood in the strandline indicates that it was not formed until the water receded after the norther. The fact that the separate lower jaw already is part of the strandline supports this supposition.

Plate 21, figure B shows another alligator gar carcass in the dorsal position perpendicular to the shore. The lower jaw has been carried off by vultures. As usual, the midline is not exactly symmetrical, but twists to one side. The carcass is two-thirds embedded in the strandline, and it is very typical that the little pieces of wood have come to lie snugly up against the new obstacle. The carcass of a somewhat smaller gar was found lying on its back completely within the strandline (pl. 21, fig. C). The lower jaw has been carried off, and the bones of the upper jaw have been somewhat pressed apart by a piece of wood lying on top of the jaw; the throat is decaying. The body has already collapsed considerably. Some of the driftwood lies in striking conformity with the axis of the carcass; another larger mass lies crossways to it.

If we study the carcasses of the small needle gar, we find again and again a peculiar curvature of the cylindrical body. It is found relatively seldom in the purely lateral position seen in plate 22, figure A. The jaws of this carcass are open, and the typical fold has formed at the dorsal fin far to the rear. The throat has been partly eaten. There is another curved individual lying next to it. The extended specimen has been deposited axially parallel to the coarse plant remains and carp carcasses. The space between the carp and the curved needle gar is filled with small twigs lying crossways; the gar's tail lies generally in the direction of the rest of the remains, following the shoreline, but then bends crossways to it. Plate 22, figure C shows a young garpike in a ventral position with the characteristic curvature. The dorsal line of the cylindrical body runs from a point on the convex side of the silhouette, in the abdominal area, to the concave side, causing the dorsal fin to lie on the concave periphery and the tail to lie lateral. In plate 22, figure D, we can see clearly the great difference between the depositional position of a teleost fish and that of a needle gar. While the former lies flat laterally, the ganoid fish lies mostly ventral, but is twisted in such a way that the line connecting the tip of the snout with the tail is distorted. The tail lies sideways and the dorsal line twists up toward the head, which lies aslant with the left side down. This curious curvature, which is doubtless intensified when putrefactive gases cause the carcass to swell after death, stresses the scale rows on the concave side and twists and stretches the connections in the skin that hold the scale rows on the convex ventral side. Plate 22, figure B shows the sharply curved carcass of a needle gar in an incomplete lateral position. The tail is purely lateral and the head purely ventral. The midline of the back runs mostly along the profile of the carcass. The lower jaw is decaying, and the tail is turned toward the shore. The ground is drying out. At the tail end of this creature, there is a piece of dried alligator skin. This characteristic collapse of the skull of a needle gar, which reminds one of so many specimens from the Solnhofen lithographic limestone, shows even more clearly in plate 22, figure E. This carcass disintegrated in water, but later dried out. Seeds that washed up are sprouting, and there are bird tracks in the mud (observed in February 1925).

The ventral position of the head clearly has more effect on the large alligator gar than on the needle gar. We see on the typically curved carcass in plate 23, figure D that the tail assumes a lateral position, the dorsal fin spreads upward, supported by its fulcral scale, and the dorsal line runs down near the concave side but remains a certain distance from the carcass profile. We see clearly how the extended convex side of the body rises gradually from the surface it lies on, while the concave side rises more steeply. Notice in the picture that a little piece of wood lies axially parallel to the tail, whereas on the convex side, most of the bits of wood lying parallel to the shore have assumed a tangential position. The jaw area is already disintegrating. This carcass dried out early and was consequently well mummified and relatively fresh looking. I want to compare it with a number of other carcasses of the same species observed on the same day, but which lay much longer in the rising water. First, we see in plate 24, figure A a fish whose position is just like that in the previous example only in mirror image: the previous carcass was curved to the left, and this one curves to the right. The lower and upper jaws have already been broken off and moved. On its back, there is a hole made by feeding vultures, probably when only this part of the fish protruded from the water. We see again that the extended outer curvature creates a gentle slope and the compressed inner curvature a steep one (see also pl. 22, fig. B), causing a twisting of the flat ventral contour and thus a displacement of the scale rows due to dissolution of the connective tissues in the skin. The scale rows themselves, owing to the way the scales are joined, are holding together better. Nevertheless, in a number of places, a good many scales and parts of scale rows have become detached, and the skin on the ventral side has completely disintegrated. We find exactly the same thing in the Kupferschiefer carcass of *Palaeoniscum* (see pl. 17, fig. C), where the dorsal part of the skin is still well attached but the individual scales and parts of scale rows are strewn all around. If disintegration continues, the detached scales will sink into the mud. Plate 23, figure A shows the carcass of an alligator gar that lay in the water for a long time; the carcass lies in a ventral position and presents a double curvature. Only the more firmly attached midline bones of the skull are still present, indicating that the head bent considerably away from the original curvature of the carcass. On one side, the scales have fallen off and have become covered with mud; on the other, the spine has been pushed sideways into the steeper part of the profile of the body. Here, too, the scale rows have disintegrated in a typical way. The twisting of the spine is relatively easy to follow.

The carcass of the large alligator gar in plate 23, figure B, whose head lay with the dorsal side down, is in a much more advanced state of disintegration. The underside of the braincase and the long parasphenoid are visible. The latter bone lies almost at right angles to the sticks and twigs, indicating the course of the strandline, and there is a bend in the neck behind the skull. From the overall position, it is clear that the fish washed up perpendicular to the shore and was bent so that the tail end is in the strandline. The place of maximum curvature of the abdomen is covered with woody debris. The convex, gently sloping side of the profile, shreds of its scaly skin preserved, is turned toward the shoreline; the steep concave part is turned away. Both the head and the tail are decaying. Individual ganoid scales are lying on the ground around it.

Finally, we come to the last stage of disintegration. We see the mud covered with large scales (pl. 23, fig. C), some lying upside down; a few are still connected naturally. The little stalks at the base of the scales make them look like arrowheads. They are still lying at the waterline, so the carcass must have decayed while moist. The whole mosaic is undergoing complete disintegration and rearrangement, and the result is extraordinarily reminiscent of what we so often see along bedding planes of the Mesozoic.

A careful look at the places where vultures have fed on carcasses reveals that they begin making holes in the back while the water is still high. (Pl. 12, fig. A and pl. 24, fig. A show this phenomenon; the first shows an alligator gar that has been moved from its original position by this feeding behavior.) Later, the opercular region is eaten and the lower jaw torn out. The scaly skin, however, is so tough that except for the parts already mentioned and the eyes, which are always missing, everything else is left either to decay or to mummify.

5. TURTLE CARCASSES AT SMITHERS LAKE

Let us now look at the remains of turtles that froze to death at Smithers Lake (pl. 25, figs. A and B). Their legs stick out so naturally and the heads, because of their weight, hang so far forward that one might think they were still alive. Figure B shows such a turtle lying in shallow water shortly after the deadly norther. Its head points toward land, as was usually true for those I saw lying perpendicular to the shore. Figure A shows another specimen in a similar position, but mummified. The extremities are completely dried out and slightly raised above the ground. The hind legs point more definitely toward the rear. Both were photographed on the same day. The difference in their state of preservation is not due to a time differential but to how far each was from the shoreline when the storm hit. A mummified soft-shelled turtle, plate 12, figure B, has already been mentioned because its neck is bent sharply backward. Plate 24, figure B shows a turtle in the infrequent dorsal position axially parallel to a teleost fish. We often see that turtles that have drifted ashore axially parallel to the strandline influence its formation; typically, their wide bodies cause the angled arrangement of subsequently arriving material. Another illustration (pl. 24, fig. C) shows a turtle in a strandline of wood; the rear edge of the shell slants upward. Plate 24, figure E shows a turtle shell embedded in a mosaic of teleost fishes with the same orientation. It is striking how seldom the voracious Chelydridae (pl. 33, figs. C and D) are found here. They are obviously either quite resistant to extremes of weather or able to burrow deep into the mud during a storm. *Chelydra serpentina,* the common snapper, whose head is covered only with soft skin, grows to 70 centimeters in length, whereas the *Macrochelys temmincki* Holbrook, which I found more often in Louisiana, grows to one meter in length and can weigh as much as 120 pounds [60 kg] (pl. 33, figs. C and D).

Beds rich in turtle remains always have a special meaning for geologists; they frequently indicate shrinking ocean basins, floodplains, mouths of rivers, or sandbars. It is typical to find turtle remains with those of crocodiles. Near Kingsville, I have seen countless dead turtles left behind after grass fires. There are a number of reports of the mass death of turtles; those of the Galápagos Islands have already been mentioned. In the sands and gravels of the Gulf Coast rivers, pieces of shell from recent, subfossil, and fossil turtles are common everywhere. The plentiful turtle remains in the Upper Jurassic at Solnhofen, Solothurn, and around Hannover are interesting; at the northern edge of the Harz Mountains, as well as in Switzerland, turtles are numerous only in certain beds. At Solothurn they are found by the hundreds in an area that can be walked around in less than a quarter of an hour, in a bed only four meters thick. Most of these turtles are lying quite naturally in a ventral position. Pieces of ribs, vertebrae, and shells are found along with complete remains.

6. ALLIGATOR CARCASSES AT SMITHERS LAKE

Alligators, although related to crocodiles, have a different appearance. They have a wide, blunt snout, which closes so that the fourth mandibular tooth is not visible. Their geographic distribution is curious. In Europe, they are found in the Eocene; in Germany whole skeletons are found, especially in the bituminous shale at Messel. Ludwig (102) has already described parts of eight skeletons, which he divided between the species *Alligator darwini* [*Diplocynodon darwini*] and *Alligator ebertsi* [*Diplocynodon ebertsi*]. Numerous remains of *Diplocynodon* have also been found recently in the central German Eocene. Coprolites and bones broken by alligator teeth are found at Messel, as well as in the Middle Eocene lignite of the Geiseltal. Today, the alligator is limited to the southeast coast of the United States, where the American alligator, *Alligator mississippiensis,* lives, and southern China, where *Alligator sinensis,* which grows to only six feet in length, lives.

The species found in the carcass assemblage at Smithers Lake is *Alligator mississippiensis* Daudin. The head is wide, the snout blunt, and the fourth mandibular tooth fits into a recess in the upper jaw and is not visible. The nostril is divided by a bony septum. This species grows to between 2 and 4.5 meters in length, the tail accounting for half of this. It can weigh as much as 500 pounds [250 kg]. The upper side is dark brown or blackish, the underside paler. In each half of the jaw there are 19–20 teeth. A clutch consists of 20–40 eggs, which need two months to incubate. This species is found from North Carolina to the Gulf of Mexico and west to the Rio Grande. It ranges up the Mississippi River as far as Rodney, Mississippi. In the United States, there is also a crocodile, *Crocodylus acutus* Cuvier, that can grow up to 7 meters in length, but usually averages only 3 to 5. It is olive-green or gray. It lives in the salt marshes of southern Florida as far north as Lake Worth and is also found in the greater Antilles and from Puerto Rico to Ecuador; it does not occur in the carcass assemblage under discussion.

Let us examine the alligators of the vast carcass assemblage at Smithers Lake. One of the largest specimens is seen in plate 26, figure C. Although the photograph was not taken until the middle of February—before then the water was so high that the carcass could not yet be seen—the creature was still lying in the water. (During the storm, the lee side of the lake was exposed to the force exerted by the masses of water being pushed southward, while at the same time the heavy precipitation raised the water level considerably. Afterward, the level went down continuously, causing large parts of the carcass assemblage to gradually dry out.) The snout points toward the middle of the lake, the tail shoreward. If a carcass like this, a tree trunk, or any other long object floats along with the waves, it becomes so aligned that it offers the least possible resistance to the movement of the water, that is, it drifts perpendicular to the crest of the wave. These objects arrive in shallow water in this position. Their heaviest and most irregularly shaped parts anchor most easily, while the parts with the least basal area are still movable. The specimen mentioned above still lies in the position in which it drifted ashore. Here and there, putrefying gases have caused the skin to burst. The legs are slack; the foreleg is in a completely passive position and points to the rear. Plate 27, figure A shows another carcass in the same position. The tail points shoreward and lies in shallower water than the head does. The body is somewhat tipped to one side due to the asymmetrical effect of the movement of the water. Both the head and the tail are bent to the same side, making the other side convex. Moving water, acting as a lever, has rotated the body somewhat on its long axis, with the result that the legs on the convex side are raised, making the side of the abdomen visible, while on the other side, the legs are buried deeply in the mud. If the waves are strong enough, the carcass can be completely rolled over at the strandline; all strandlines, which are unconsolidated when first deposited, become tightly compacted by this process. Plate 27, figure B shows an alligator in the dorsal position whose head is already bent around parallel to the strandline. The tail, however, points somewhat away from the shore, obviously because it lies in deeper water. In this position, the alligator does not conform exactly to the course of the strandline, but the pieces of wood on top of it do and are positioned like chords to the bow of the carcass.

Interference phenomena in turbulent water—the impact of waves on objects lying at an angle to them—create a kind of transitional zone of deposition; carcasses deposited there show such bending and bidirectionality. The carcass in plate 27, figure C is an example. The tail goes in a completely different direction from that of the head. The legs diverge sharply on the convex side, and they seem to be farther apart than those on the concave side. The convex half of the body is also pressed deeper into the mud than the concave side: this carcass, too, has been turned on its long axis. The inconsistent movement of the water is also reflected in the arrangement of the pieces of wood that came ashore with the carcass. One piece lies axially parallel to the tail; another lies across the hind leg on the opposite side of the body and is oriented exactly the same as the first piece. Yet another, bigger piece lies adjacent to the one just mentioned at an angle of fifty degrees, and the head of the carcass bends in the same direction as this third piece.

A larger carcass is seen in plate 28, figure B. It has been rolled slightly on its long axis; the tail points toward the shore, and the snout toward the water. The slight twisting is best seen in the fold of the skin on the right side of the throat. Plate 28, figure C shows us the same animal somewhat later, on March 15, 1925. The water has totally disappeared. The previously described position of the animal—perpendicular to the shoreline—is the result of its position in relation to a strandline of bits of wood formed when the water level was lower. The position of the legs on the right side of the body corresponds to the strandline. The carcass has already decayed considerably and lies flat. The slight twisting on the axis is, however, still recognizable, because the bony plates on the back extend to the left slope of the body and the skin of the abdomen is visible on the right side. We also see the fold in the skin on the right side of the throat where the large muscles that close the jaw lie. The asymmetrical position is further reflected in the orientation of the legs: those on the right are extended and somewhat splayed, while those on the left are flexed. The forefoot points forward, the hind foot rearward. In the flow shadow of the left forefoot, seeds that have been washed up are sprouting.

The photograph in plate 28, figure A was also taken on March 15, 1925. The carcass of this alligator lies perpendicular to the shore, and the position of the legs very much resembles that shown in plate 28, figures B and C. The snout and forelegs lay in the water longer than the other end and are considerably macerated; the tail, however, is mummified. The characteristic bending of the tail is caused by the fact that from the place where two lateral rows of bony plates come together to form a single bony crest, the tail is flattened and lies on its side. The carcass in plate 29, figure B is of an animal that died on the same day as the previous example and was also photographed on March 15, 1925. The damp ground indicates that the water receded from this area only very recently. The tail is better preserved and mummified, and there is a characteristic twist at the end of the vertebral column. The skull has been moved away by scavengers, but the lower jaw and the broken-off upper jaw are still connected to the carcass by shreds of skin. The carcass is lying on its back. The femur of the left leg as well as a large number of ribs are clearly visible. The pectoral girdle is considerably dislocated. Because the posterior end of the carcass is somewhat less decayed, large parts of the skin are still present, but all that is left of the large alligator shown in plate 29, figure A is the skeleton. It lies perpendicular to the shore, and the spine has the typical curvature because the head lies dorsal and the tail lateral. The middle part of the spine lies partly dorsal, making the curvature of the neck completely comprehensible. Because of this crooked position, one-half of the pelvis has sunk into the mud, while the other lies on the surface. The right humerus is pushed far forward. Both valves of a large freshwater clam *(Anodonta)*, the convex sides uppermost, are lying opposite the pelvis, which is already partially buried in the mud. A few little pieces of wood are lying across the curvature of the neck, forming a chord to the arc of the neck.

Let us go back to the alligator carcasses found in the strandline: plate 30, figure D shows driftwood lying axially parallel to the strandline and an alligator carcass in the dorsal position at somewhat of an angle to it; plate 29, figure C shows the same thing, but with ganoid fishes. Both photographs were taken on February 25, 1925. We see clearly that the oncoming waves have shoved the unconsolidated debris together so tightly that the resulting mass put a stop to the forces that had caused the carcass to turn on its long axis.

Plate 30, figure B shows an alligator carcass in a strandline with the typical angled structure. The alignment of material coming ashore rotated almost ninety degrees after the carcass washed up, so that the axis of the body of a needle gar, washed ashore later, lies right across the tail of the alligator. In the middle of February, the alligator carcass was still fairly swollen by putrefying gases. Plate 30, figure C shows the same carcass a month later. The position of the sticks has been somewhat disturbed, partly by feeding vultures but mostly by the complete collapse of the now partially mummified carcass.

The smallest alligator carcasses had already been washed ashore before the firm, orderly strandlines formed. Plate 30, figure D shows such an isolated carcass. The end of the tail of this young alligator lies laterally, and the bend is forming at the typical place mentioned above. Both hind legs and one foreleg point slackly to the rear. Vultures have eaten the eyes. A needle gar lies at an obtuse angle to this alligator. The head of the fish, with part of the jaws decomposing, is lying on top of sticks whose position corresponds to the axis of the alligator carcass. This situation is the commonly seen result of two intersecting currents. The whole surface is covered with seedlings because the animals are lying in a place where seeds were washed ashore.

Plate 30, figure D was taken in the middle of February 1925. If we look at these somewhat dispersed carcasses one month later, we see that because they dried out quickly, they are much better preserved than the completely decomposed carcasses of animals that lay almost the whole time—but not even a quarter of a year—in water. The lower jaw of one carcass (pl. 31, fig. E) has been carried off, and the tail has disintegrated extensively. The head lies dorsal, the tail lateral. The neck shows the typical curvature, and the legs on the concave side of the torso are extended; on the convex side, the foreleg reaches over the torso, while the hind leg is more flexed. On the convex side, we see the outer row of bony dorsal plates; on the concave side, the skin covering the abdomen has shriveled.

Plate 31, figure C shows an almost completely mummified animal in the dorsal position; on the concave side, the skin covering the abdomen has swelled and burst, forming a convex curvature. The animal in plate 31, figure D is in a similar position. Plate 31, figure B shows a desiccated carcass in the ventral position with a sharp S curvature in the vertebral column. Such pronounced drying-out of a carcass probably happens only when it is quickly carried to a dry place and exposed to the strong rays of the sun. Plate 31, figure A shows another example—an almost circular carcass of a young alligator lying next to a turtle. Its position is reminiscent of the *Protorosaurus* specimens of the Mansfeld Kupferschiefer and the small *Pelagosaurus* that Pompeckj collected at Holzmaden for the Geological Institute at Göttingen. The tip of the tail does not, however, point toward the snout, but backward, forming an S shape. The skin of this small alligator has dried out and sticks tightly to the bones. Vegetation is beginning to take over. Such places must have once been pools that formed amid washed-up debris; floating seeds came to rest only at the peripheries of such pools.

In conclusion, I give an example of how to recognize pressure from gas in alligator carcasses (pl. 33, fig. A). The individual lies on its back; the legs are first raised above the pectoral and pelvic girdles and then stick out perpendicular to the body. The cloaca is everted, and the skin between the plates is clearly visible, especially on the extended left side of the neck. The degree of decomposition of a vertebrate carcass at the time of burial is of great significance; many different stages are possible.

The information given in Chapter 4 of this book will be amplified in the next work by this author: Ganoidfischleichen im Kupferschiefer und in der Gegenwart [*Palaeobiologica* 1: 323–56].

5   Carcass Assemblages and Concentrations in the Geologic Past

Did natural processes in the geologic past cause masses of carcasses to accumulate, as happened recently at Smithers Lake? There are quite a few distinctive examples apart from the more frequently observed deposits of mammal remains. Somewhat problematic is the famous group of *Aetosaurus ferratus* O. Fraas in the natural history collection in Stuttgart [Staatliches Museum für Naturkunde]. That group has usually been explained by saying that the animals were surprised while still alive by a collapsing dune or water-borne sand and that their positions with curved spines clearly betray their final agony. This explanation does not seem so evident when we consider a passage from the study by Fraas (50) that has usually gone unnoticed, and which, therefore, I quote:

> When setting up the exhibit in the Museum of the Royal Natural History Collection, in order to save precious space, empty rock was removed from between pieces 7 and 8 and pieces 10 and 11, and the joined pieces 20 and 21, so that the size of the group is now 1.43 square meters. But this procedure changed nothing in either the position or the arrangement of the individual pieces.

As far as researchers are concerned, these measures are exceptionally regrettable. The twenty-four individuals originally lay on a surface about two meters square. They are not lying simply axially parallel, although some do, but in a tangle, in four main directions. One could easily form the impression that the animals, stunned or killed by the same force, were brought together in quick succession, perhaps adjacent to an obstacle that slowed down or divided flowing water in such a way that the carcasses could drift up from two sides. The bodies of the animals lay in a depression formed where a swelling lens of sandy marl came into contact with a sandstone bank. Fraas says further: "They lay there together in an old pool in the sand just as they were washed up, the bodies arranged haphazardly according to the water currents."

Based on the advanced state of maceration of the carcasses, Fraas himself concludes that it is unthinkable that evidence of the death agony can be seen. While he is of the opinion that the natural grouping of the remains indicates they were washed into a pool, I prefer to think of a strandline, where the sand was churned by strong wave action and then, in a manner of speaking, armor-clad by the carcasses washed ashore.

A peculiar kind of concentration is also shown in the position of the group of *Koikilosaurus coburgensis* described by Huene (73). The skeletons of three specimens are lying close together; the axes of their bodies are absolutely parallel. The same position is seen in the group of *Diplocaulus broili* from the Texas Permian, described by Broili (27). In this instance, there are also three individuals lying close together, axially parallel, with even a few matching, angular displacements of the spine; they deviate from the main direction but run parallel to each other. The unusually wide, flattened skulls with their widely flared rear corners are very irregular in shape and consequently lie head to toe—the one in the middle heads in the opposite direction from the other two. These tadpolelike amphibians, whose long bodies and unusually short front and hind legs hint strongly at an aquatic way of life, must also have died from a common cause, their remains then concentrating post-mortem by drifting ashore. The two splendid groups of *Neusticosaurus pygmaeus* and *Neusticosaurus pusillus* from the Lettenkohle at Eggolsheim, pictures of which I publish with the kind permission of Dr. Berkhemer, curator of the natural history collection in Stuttgart, show this phenomenon even more clearly (pl. 32, figs. A and B).

The group of nine specimens (pl. 32, fig. A) of *Neusticosaurus pygmaeus* from a dolomitic bed in the Lettenkohle at Eggolsheim served as the type specimen of this genus of saurians for Eberhard Fraas, but no complete picture of them has heretofore been published. The slab shows nine individuals that were clearly part of a margin deposit; they came together in an orderly way following a die-off of catastrophic proportions. The body axes of five of the individuals are axially parallel to each other. The rest are parallel to each other, but the group forms an angle of about 35 degrees with the first group. The first five were obviously deposited before the other four. In the first group of five, the head and tail of the first, outermost specimen is bent around away from the group. The next specimen is incompletely preserved; its tail, pelvic girdle, and the lower part of its body do not extend beyond the first individual. Both the neck and the tail of the third bend sharply in the same direction. The fourth specimen is turned 180 degrees; its neck and head are also bent to the same side, and cross the tail of the third specimen, pointing toward the incomplete tail of the first. This turning around from head to tail is surely no accident, but rather the effect of the current on different body parts, first as they floated and then as they trailed around the center of gravity. The fifth specimen is the largest of all. The end of its neck reaches as far as the front legs of the first specimen. The upper right front leg lies parallel to and even with the right hind leg of the fourth specimen and at the same angle to the axis. Depending on how splayed the legs are, the sixth, seventh, and eighth specimens have come to lie close together at an angle of about 30 degrees to the others. (This change of angle is also evident in the formation of strand-lines composed of sticks and twigs; in such shore-parallel deposits, the angles gradually change.) The outermost specimen is curved. Four specimens are lying so that the ends of their tails are turned toward each other.

Another slab from Eggolsheim with closely related remains (pl. 32, fig. B) is the basis of E. Fraas's description of *Neusticosaurus pusillus*. It, too, comes from the Lettenkohle at Eggolsheim, and the individuals on it are seen from the ventral side, exactly as are those on the other slab. First, there are two fairly incomplete individuals lying axially parallel but head to tail; the leg of one bends up by the neck of the other. The neck of another specimen curves around close to the somewhat splayed leg of one previously mentioned and lies perpendicular to the axis of the latter's body. The axis of this third specimen deviates only 35 degrees from the axis of the second, but its cervical spine is so sharply bent that it lies perpendicular to the bodies of the first two individuals. The caudal spine of a fourth specimen lies symmetrical and parallel to the neck and head of the third. Here, too, we see the angled structure and alteration of the leading edge of a strandline formed automatically by the addition of new material. There is a fifth specimen lying almost parallel to the third—also at an angle of about 35 degrees to it. A few others are lying parallel to the cervical spine with the right-angle bend. There are roughly three different axes, which form angles of 35, 80, and 40 degrees with each other. This, too, is typical for a strandline when the components are long and somewhat irregular in shape.

One of the neusticosaurians on the first slab shows a bending around of the neck parallel to the body axis; this position probably did not occur through mummification, but rather through the folding around by the current of the long, flexible neck at the point where it was most resistant to the force of the water. It is the same kind of bending we have seen before. The same position can be seen in the excellently preserved *Anarosaurus pumilus [sic]* Jaekel [*sic*] [*Anarosaurus pumilio* Dames], closely related to *Neusticosaurus* from the Wellenkalk at Remkersleben [West Germany] (in the Göttingen collection). The impression that this specimen was washed up near shore is reinforced by the finding in the same locality of well-preserved conifer remains, which Schütze (146), who gave the original material to the natural history collection in Stuttgart, has written about.

The skeletons of *Pantelosaurus saxonicus* von Huene (77) [*Haptodus baylei*] (see fig. 28) also form a strandline:

> The six skeletons come from the Queen Carola Mine near Döhlen [East Germany] in the Plauenscher Grund. The bed lies in the Lower Rotliegend immediately above bituminous coal seam 1 in the so-called green bowl, a layer of greenish gray mudstone only eight centimeters thick at that point. Above it, right on top of the skeletons, there is a loaf-shaped mass of mudstone thirty to seventy centimeters thick, with indistinct plant remains.... It seems to have been a brothlike mass of mud that flowed over the saurians.

The skeletons themselves clearly form a strandline. Four individuals form a firm foundation against which the fifth has come to lie snugly. The position of the first, which lies across the fourth, is similar to that of the second and third. The first is a complete skeleton, slightly bent in an S shape from head to tail. The second is complete from head to pelvis and the scapula is twisted sideways; the snout lies underneath the first one. From the head to torso, the third lies parallel to the torso and tail of the first. Its left front foot is lying on the snout of the first. The skull of the fourth, showing its underside, is lying next to the pelvis of the first and on the left side of the third. The torso curves in a half circle around the front part of the torso of the first one, and the tail forms a loop between the second and third. The skeleton of the fifth is the most complete and goes past the skulls of the fourth, first, and second. Its skull is almost covered by the neck of the sixth, and largest, individual, which is represented only by its skull, neck, and the beginning of its torso.

A similar example of the addition of new remains to some already in place is found in the prime specimen of *Sclerocephalus haeuseri* Goldfuss; behind the skull, in the sticky clay that enveloped it, lay two skulls of *Branchiosaurus amblystomus* juv. Credner [*Apateon*]. Ammon (13) explains this proximity by saying that the *Sclerocephalus* fed on the *Branchiosaurus*. This example is just like the one in which a skeleton of "*Aphelosaurus*" Jaekel [*Weigeltisaurus*] lies close beneath the fins of a *Coelacanthus*, in the Kupferschiefer.

Peterson (126) described the excavation of fossil mammals from the Miocene of Running Water, in Sioux County, Nebraska. The lighter color of the twenty-inch-thick bone bed stands out against the tan sandstone. There were fifty to sixty skulls, lower jaws, and other remains of *Dinohyus* and *Moropus* found on a surface of five square feet. Abel (7) has recently written on the same subject. He describes how the Lower Harrison beds near Agate, Nebraska, are strewn with numerous remains of fossil mammals, mainly *Diceratherium,* which are no longer naturally articulated. In the Colorado Museum [Natural History Museum of the City of Denver], there is a very typical block 1.65 × 2 meters in dimension, which Abel (7) shows in his illustration 240. On this surface there are twenty-two skulls and many other bones—an estimated 4,356 pieces in all. According to Mathew (Abel 7), as of the present, only about one-twentieth of the bones have been excavated; he estimates that as many as 16,400 *Diceratherium,* 500 *Moropus,* and 100 *Dinohyus* individuals are there. The abundance of bones from the Pliocene of Maragheh in Persia [Iran] is legendary. In the volcanic masses on the western side of Lake Urmia, just as in Nebraska, there are "nests" of bones, predominantly of single species, that have been reworked by rivers. *Hipparion* is found almost everywhere, rhinoceroses and even-toed ungulates are usually found separately, and predators usually lie with the ruminants. Both fossil-rich beds are interrupted by sandy material and are 0.6–1.0 meter thick.

The savannahs of the earth, naturally treeless or sparsely wooded, are rich in accumulations of bones; the pampas in Argentina and the loess in China are two good examples. Significant bone accumulations are also found in the Russian steppes, especially along the banks of the Volga and its tributaries, or near Odessa and Nerubaj, in southern Russia. In such places, mass death is easily caused by dust- and snowstorms and cold, which even today devastate the mountain livestock of the Caucasus, and by drought, epidemic and flooding.

Cold north winds or snowstorms are not as yet implicated in the vertebrate accumulations in Nebraska, and yet they are among the most devasting events we can imagine. The carcass field at Smithers Lake owes its origin to a single norther, which killed cattle not only there but at other localities as well, around 1.25 million head in all. If we turn to Eurasian examples, we find exactly the same thing. The number of animal victims of a single snowstorm in the Russian and Siberian steppes is often enormous. According to Helmersen, an unexpected, sudden, fast-breaking snowstorm in the district of Saratow, in 1832, killed 10,500 camels, 280,000 horses, 30,480 cattle, and 1,012,000 sheep. Crazy with fear, the animals had crowded together, and the result was vast carcass assemblages. Similar reports come from many travelers in Asia, and quite a few fossil bone accumulations in Russia and central Asia must be attributable to such winter storms.[27]

27. Worsening weather often increases the possibility that a carcass that might have decayed in the warmth of summer is preserved, first, because colder temperatures arrest putrefaction, and second, because the earth is softer for longer periods of time and assimilates remains easier. In the years between 1882 and 1884, on the New Siberian Islands Ljachov, Kotelny, and Faddejev, Bunge gathered thirty-five hundred mammoth teeth, even though teeth had been gathered there for over one hundred years. Pfizenmayer has described how at the beginning of summer, just after the snow has melted, the swelling rivers undercut their banks, which then collapse, releasing mammoth tusks and rhinoceros horns. As late as 1908, the Wrangel expedition found an extraordinary abundance of fossil animal remains in areas that were literally saturated with mammalian and other bones. A large part of the Siberian finds has long been sent to China as dragon teeth. In

(Continued)

Grottos, caves, niches, and dolines also serve as collectors of vertebrate remains. From a sinkhole twelve meters deep and five meters wide near Hundsheim [Austria] (see Schaffer 138), in addition to an almost complete rhinoceros skeleton, the bones of elephant, swine, roe deer, deer, goat, wild sheep, bison, hyena, wolf, jackal, bear, saber-toothed tiger, leopard, wildcat, ferret, weasel, porcupine, hare, dormouse, water vole, bat, hedgehog, mole, shrew, black grouse, swallow, thrush, goshawk, snakes, lizards, and frogs were excavated. Hamilton (61) reported on remains of extinct birds found near Ngapara, New Zealand. The substratum of a limestone cliff wall was giving way, causing continuous spalling of the

---

1821, a single Yakut Indian brought five hundred pud [an old Russian measure equal to 16.38 kg] from the New Siberian Islands to market. In the 1840s, Middendorf estimated the yearly production of fossil ivory in Siberia at 120,000 pounds. In 1873, 1,140 Siberian tusks came onto the London market. Great quantities were sold at wholesale to the Japanese. Pfizenmayer estimates the total average annual production in northeastern Siberia at 32,000 kilograms.

We can assume that in the area of the New Siberian Islands, whole herds of mammoths perished in snowstorms and as a result of other natural events. According to reports by E. Hennig (*"Kentrurosaurus aethiopicus:* Die Stegosaurier-Funde vom Tendaguru, Deutsch-Ostafrika," *Palaeontographica,* supplement, vol 7, 1(1), 1925), the kentrurosaurs from the Tendaguru beds of German East Africa [Tanzania] also died in herds, young and old animals together. The main locality contains the thoroughly jumbled elements from about fifty individuals of all ages. Hennig writes the following:

> Sometimes, when we thought we had found remains of one skeleton, even in the smaller excavations where there are usually a couple of ilia, a couple of sacra, and a long and a shorter leg bone of the same kind, we turned out to be wrong. And what is even odder is that it is thought that a kind of sorting of certain skeletal elements could be observed: in one place, several of the sacra were found close together; in another, a number of leg bones; in yet another, especially numerous vertebrae of all kinds. Pieces of front and hind feet, which are almost totally lacking in the large herds of the excavation sites, were found in masses in another place. . . . Surely, however, we can assume that this group of more than fifty stegosaurs did not accumulate one after another by accident; what we have here is a herd—not exactly a typical phenomenon for reptiles. Very young animals, half the size of typical adults, are lying underneath. Their age, of course, cannot be determined.

There is no grouping of the animals according to age, but rather a progression from the youngest to the oldest.

> Where such young animals are killed in such large numbers, we assume they must have been taken by surprise and overcome while still alive. It is no "dying ground" in the usual sense. Where whole herds are buried, no substantial transport (aside from reworking and local flushing by water) can be considered, either. At this site, the cause of death assumed catastrophic proportions and turned the habitat into a grave. Nevertheless, it should be emphasized again that there was time for decomposition to take place, for extremities such as the skull and legs to rot off, and for skeletal connections to

(Continued)

wall; between the cliff and a more or less complete section of material that had broken off, earth, stone, and decaying vegetation had accumulated along with quite a few remains of extinct birds such as *Notornis, Fulica, Aptornis, Cnemiornis, Harpagornis, Carpophaga, Ocydrornis* [*Ocydromus*], and *Anas fischii* [*Euryanas finschi*]. These remains are also found in fissures on the limestone plateau above. The birds probably found their way into these cracks when the vegetation was denser; the cracks filled with debris and bird remains, and as the wall collapsed further, the remains were strewn about. In the Brazilian bone caves, remains of monkeys, bats, carnivores, rodents, and marsupials were found.

Abel and his students have reported on the abundance of cave bear remains in the Mixnitz caves. In the Dragon cave near Mixnitz, in the Steiermark [Austria], about fifteen million kilos of pure phosphate has accumulated. Manure piles up quickly in such an enclosed space. In America, bat towers have been built in swampy areas to combat malarial mosquitoes; bat excrement accumulates on the floors and is used as fertilizer. In Greenland, accumulations of walrus droppings have been observed.

Fissures in the rocks of the Swiss Jura and the Rauhe Alb, the Dalmatian Karst, and the caves of southern France form well-known vertebrate localities. The geological occurrences of fossil brids, often connected with particular formations, are interesting. Near the mouth of the Thames lies the Island of Sheppey [Great Britain]. Here, in the Eocene, numerous fishes, turtles, crocodiles, palms, conifers, and many tropical deciduous trees along with bird remains are found. The rich bird finds from the Montmartre gypsum in the Paris Basin, a locality just as rich in mammal remains, are of the same age. Until now, the richest locality for Eocene birds has been the Quercy phosphorites. In the Messel bituminous shale a snipe was found, and recently excavations carried out by Johannes Walther in the Eocene lignite of the Geiseltal have also produced birds. The Steinheim Basin, Sansan [France], and Pikermi [Greece] are Late Tertiary bird localities, and each has its special aspect. The bird world of

---

dissolve before final burial. Currents, waves, or scavengers may have been responsible for the lamentable confusion of the bones.

Valuable observations on concentrations of bird remains were made by Wetzel in Chile ("Vogelmummien und Guano in chilenischen Salpeterablagerungen," *Cbl. für Mineralogie,* etc., 1925, p. 284). In the brown, phosphate-rich sand and conglomerate lenses found in the nitrate vein, which contain numerous bird remains, he saw nesting colonies of the storm petrel covered by alluvial talus cones caused by sudden deluges. There are thin beds with many bones, parts of feathers, droppings, and remains of food. At other places, he found granitic detritus transported by rivers and cemented with salt, containing a great number of bird mummies. At one time, when the climate there was still not so extreme, marine brids had their breeding places in the mostly dried-up riverbeds. Wetzel emphasized correctly the catastrophic nature of the destruction.

the Pleistocene has come down to us mainly through the California tarpits.[28]

The purity of the bone fringe formed when concentrations of carcasses wash up is often striking. Frequently, however, clay balls betray the fact that a carcass field was first subject to processing by water—a washing-away of the lighter bones, leaving the heavier remains behind to be completely reworked or totally dispersed; for this to happen, a simple rise in water level, which often appears as a transgression, is sufficient. It is of critical significance how far decay and disintegration have advanced before this happens. Just think how differently dried-out, mummified, curved carcasses behave in a strandline compared to individuals that decayed in water. There are also strandlinelike accumulations of bird feathers, in fact, in the Nördlinger Ries [southern Germany]. According to Ammon (13), most of the Upper Miocene lacustrine limestone contains feathers of *Anas risgoviensis* Ammon in abundance.

A species of animal can inhabit a water basin for a long time without leaving remains behind in the basin sediments to prove its presence. But if an especially unfavorable year comes along; if there is any sort of disturbance during the breeding period; if too much sediment is carried in or if the basin dries out; if one generation of a particular population, because of special weather conditions, increases sharply and unexpectedly, and then dies when the favorable conditions do not last, a whole bedding plane can be covered with the remains of the affected organisms. Within a restricted habitat, such a die-off can be catastrophic, but for the species it usually means only anastrophe. An example is the occurrence of *Branchiosaurus* [*Apateon*] from Gottlob, near Friedrichroda. There, new excavations conducted by Kellner have removed the overlying hard rock from the surface of a large area of the plant-producing shales. In the process, a whole series of remains of these amphibians has again been found; however, the creatures are not dispersed throughout the whole shale bed, but lie mostly on a certain bedding plane. They all lived at one time and died together. It is possible that the whole population of the local habitat lies together in this one carcass assemblage. In this instance, the concentration is not a strandline, but an even distribution of the remains over the surface.

After death, vertebrate carcasses are often subjected to the same mechanical laws of transport by geologic agents as any other component of the sediment. Their specific gravity, which changes according to changing

28. Simmons ("Sinbads of Science," *National Geographic Magazine,* vol. 52, no. 1, July 1927, illustration p. 27) has just given an instructive, detailed picture of the sandstone surface of the uninhabited Cape Verde Island of Cima. The surface occupies a plateau that supposedly has served as a dying ground for storm petrels for a very long time. In any event, the whole surface is covered with a thick layer of vertebrae and hollow bones from these birds. The restricted size of this area, compared to the wide expanse of water where the birds feed, gives rise to such a concentration of bones.

buoyancy, and the relation of mass to surface area specify the mechanical arrangement, and after sedimentation, when all the differences have been evened out, the various elements are in equilibrium. If the carcass falls apart because of decay before it is buried, a "disorderly" pile of debris is formed. Parts can be carried off not only by scavenging vertebrates but also, to an extraordinary extent, by crabs. Then, if the water is in continous motion, no matter how slight, the individual elements of the skeleton are arranged mechanically, often developing an alignment that is axially parallel (see fig. 8). The same phenomenon causes schools of living jellyfish and masses of vegetable detritus to gather in strips; it also accounts for the bands of *Coeloceras* in the Württemberg Posidonia Shale. A simple form of the transformation is the arrangement of submerged carcasses or those washed ashore into a deposit with a certain orientation.

Jaekel (79) worked on the question of the distribution and accumulation of fish remains in the upper Muschelkalk of Alsace-Lorraine [France] and concluded that the greatest accumulation is found in the upper *semipartitus* zone, and that layers with fish remains recur and can be observed as far as the dolomite boundary. More recently, Deecke (40) explored the origin of these bone beds in great detail. He maintains that we can expect to find whole fish skeletons only in still water and at places where there were no other animals to devour the carcasses. The so-called fish shales were formed in such places—quiet lagoons and bays where the animals were buried in toto in the bituminous, sapropel-like, muddy sediments. The situation seldom occurred in German Triassic seas; it is somewhat more frequently found in the Buntsandstein. Deecke explains the accumulations of *Semionotus* by saying that ponds either filled with airborne debris or dried up, and the fish burrowed into the mud and then died. He says that the fish found in the upper Buntsandstein were washed-up carcasses that had been buried under masses of sand on the flat coasts. Deecke also studied in detail the bone beds that are so significant for Germanic facies. These facies extend from the Wellenkalk to the Rhaetian, sometimes only as thin layers strewn with ganoid scales, sometimes as thicker beds studded with all kinds of bones and teeth. I see in them the result of shallow water where the effects of the waves extended to the bottom, and the smaller, harder body parts, less subject to decay, spread out in a layer, the bodies themselves having been completely destroyed. The bone bed continued to grow intermittently as the quantity of water diminished, as happened in the Röt, in the *orbicularis* zone of the upper Muschelkalk and the *trigonodus* dolomite, and in the border dolomite of the lower Keuper. In that dolomite, which terminates the marine sequence, as well as in the *orbicularis* beds, clam shells (*Myophoria orbicularis, M. goldfussi, Gervillia costata, G. socialis*) were swept into compact piles. In addition, within the blocked German Triassic sea, there may have been a mass death in the upper Lettenkohle due to the change to an arid climate. A few forms were either preserved or lived at the edges of the basin in the mouths of rivers, where life was still possible, and migrated back into the basin when it filled, only to die off when the water evaporated again. This is how I would account for the deposits of fish scales in the Gipskeuper and Steinmergelkeuper strata of Baden and Württemberg.

In my view, however, the Rhaetic bone bed has an entirely different explanation. During that time, there was doubtless a transgression of the sea over the whole wide German Triassic area, and so the explanation cannot be based on a lack of water. Therefore, I compare this bone bed with the Tertiary Lamna, bearing deposits of similar origin, which have exactly the same massive, even appearance. Numerous shark remains, ray teeth, otoliths, and so forth are found in the Baden Miocene and in other Tertiary beds at the base of the transgression series. At Lake Constance [West Germany], in the lowest marine Miocene, there is a band of such bones a handbreadth wide, exactly as in the Rhaetic [Rhaeto-Liassic] bone bed in Württemberg. At the border between the Cretaceous and the Paleocene at Schonen, Denmark, and as glacial drift in Pommerania [Poland], there are breccias packed with shark teeth. Where the Gault and Cenomanian expand in northeast Germany, the same thing is evident. In none of these examples can it be an instance of erosion of older, previously deposited remains. In the shallow water near the coasts, there must have been a rich fish fauna whose remains were incorporated into such typical shore sediments. Crocodiles, labyrinthodonts, and dinosaurs lived in these areas of the Rhaetic, too. Their bones are also found in the bone bed; the small "marsupials" may have lived on carcasses washed ashore. From this standpoint, the Rhaetic bone bed should really be included in the Lower Lias, which the French do sometimes when they call the whole Rhaetic "Infralias." Therefore, it is noteworthy that Erni recently described the continuation of *Ceratodus parvus, Sargodon tomicus,* and *Acrodus minimus,* that is, the typical Rhaetic bone bed forms, on into the *angulatus* zone of the Lias (*Centralbl. Min.* etc., 1926; 241–253).

It seems to me that most [aquatic] bone beds originate in lagoons behind sand pits; these lagoons are often one hundred kilometers or more long and have only a few inlets or entrances. The ebb and flow of the tide or the compression of the water by the wind causes relatively large quantities of water to pass through these narrow passages. When the volume of moving water is thus compressed, its speed increases and, as a consequence, currents continuously sweep the floors of these lagoons. Accumulations of fish remains are more likely to be formed as a result of primary deposition in areas of sedimentation where the speed of the water currents is such that only the bones and not smaller particles are allowed to settle; it is less likely that fish accumulations are formed as a result of previously buried remains being washed out of their sediments and redeposited elsewhere. Brackish-water lagoons on sinking coastlines have unusually large fish populations, made up largely of sharks and rays, but these populations are usually not rich in species. Large schools of fishes also show up periodically in these lagoons, and their natural deaths after breeding can occur in vast numbers. At ebb tide, many fish fall victim to birds. Plate 33, figure E shows a muddy lagoon area east of Matagorda Bay, near Sargent, at ebb tide. The water level is marked by a fine, white strandline composed of small elements: fish scales and bones, and remains of crabs. Up to this line, in the foreground, there are no bird drop-pings because they have been processed by the water and added to the trashline. However, the surface behind the line is marbled with white droppings. We notice than that a single ebb and flow can create linear concentrations of fish remains. Many examples of estaurine sediments are found not only in the Tertiary but also in the Carboniferous. In the region around Edinburgh [Scotland], Traquair found the estuarine fish fauna in the Upper Carboniferous to be completely different from that of the Lower Carboniferous.

Reports of enormous fish kills are plentiful: Almost every ten years, a red mite appears in Walvis Bay [South Africa] and causes a fish kill of such magnitude that the water is said to be completely covered with dead fish. Poisoning by minerals and gases have likewise often been observed. Another example is the massive die-off of fish during the iron "bloom" in Siberian rivers, which Baron von Toll has reported in detail. In west Texas and in many other areas, when the water level in the rivers is low, there is a bloom of algae, and vast numbers of fish die from carbon dioxide poison-ing. Changes in salinity very frequently cause fish to die. Freyberg (55) ob-served in Mar Chiquita [Argentina] (a basin of concentrated salt solution into which, however, a completely isolated source of fresh water emptied) large accumulations of fish that had died and been preserved by the salt solution and had then floated ashore, where they were mummified. Freshwater fish in coastal lakes die off when there is an ingress of salt water. The most well-known instance occurred when the narrow neck of land that separated the Limfjord from the ocean to the west broke in 1825. After a storm tide, salt water entered and killed all the freshwater fish, which floated ashore in unbelievable numbers; some were even buried along with eelgrass beneath the sand carried ashore by the storm tide.

The reverse can also happen: a flood of fresh water can cause a catas-trophic die-off of marine fish. When waters from the Everglades, in east-ern Florida, break out into the ocean, as sometimes happens, the coast-lines are carpeted with dead fish. These devastating events are connected to the extensive karst network occurring in the bedrock of the Florida peninsula, formed from a comminuted limestone of organic origin. Rivers disappear, cave systems fill with water, and huge springs of great clarity and depth appear, all of which testify to an unusually abundant supply of water, as do the even more numerous lakes and swamps. These bodies of water are connected to and in part caused by the dissolution of the lime-stone, and other karst features, such as the collapsing of the roofs of underground, water-filled cave systems. Lake Okeechobee, with an area of twenty-six hundred square kilometers and a depth of only six meters, is one of these "lime sinks." This huge lake is connected with the Everglades by underground canals. Only under the influence of easterly storms does it drain westward into the Caloosahatchee River.

The Everglades, however, drain into the Atlantic Ocean through the Miami River as well as into the Gulf of Mexico. The direction changes according to the season and the velocity of the prevailing winds. This complex relationship makes the catastrophic breaking-out of Everglade waters understandable.

Andrée (9) reports on the natural mass death of the fish *Mallotus villosus* Müller, which occurs annually after the breeding season and which, according to A. Jort, regularly coincides with radical temperature changes in the Barents Sea. The floors of calm bays and inlets of the harbors at Disko and Lodden [Greenland] can be completely covered with fish carcasses. Remains of *Mallotus villosus* form the growth centers of concretions known as *marlekor* found in the late glacial, polar sea clays in Greenland and Norway.

The bottom of the Bay of Callao [Peru] is black and smelly. There are sometimes enormous die-offs there of the fish on which the cormorants of the islands of northern Peru are dependent. The border between warm and cold currents is fairly unstable. The Humboldt current, which near the coast has a temperature of 60° F (16° C) and on its outer side of 78–81° F (20–21° C), flows along the coast of South America from southeast to northwest for a distance of three to four thousand nautical miles. A warm current coming down from the north meets it. Because the Humboldt Current is considerably cooler than befits the geographic latitude, it is much richer in organisms than the warmer current; great schools of anchovies and other small fish swim at its surface. Multitudes of *guanayes* (cormorants) depend on them for sustenance. Because the islands where the birds live have a desert climate, the guano, important as fertilizer even in prehistoric times, is preserved. The annual yield has grown from 25,000 to 90,000 tons, of which Peru itself uses 70,000. The guanaye, an Arctic form, accompanies the cold current as far as six degrees south of the equator. The place where these cold and warm streams meet is a rich but unstable environment, as has already been indicated, and in fact, severe, even catastrophic die-offs in these areas have been reported repeatedly; those in recent times have mostly been at the expense of the cold-water fauna.

In 1897, Kissling reported on a swath of dead fish twenty-five hundred meters long and as much as five meters deep. Walther (170) reported that in the Volga Delta, great schools of freshwater fish were carried by currents into the salt water of the lagoons on the Caspian Sea and died there. There, the carcasses were partially mummified after being washed ashore. Whole schools of fish can also swim ashore and become stranded.

Such an event accounts more than anything else for the schools of *Leptolepis* [*Leptolepides* and *Tharsis*] of the Solnhofen lithographic limestone. Erich Thomas (164) observed on the Gulf of Mexico, on the Mexican coast near Fenixtepec Bar, some fifty kilometers south of Tuxampa, the unusual but not unique instance of an estuary that has no exit at all from its lagoon. A sandbar separates the Rio Fenixtepec from the Gulf of Mexico, so the debouching of the fresh water can only take place underground, by seepage through the sand. When the wind is high, large quantities of water from the gulf are thrown over the bar. The water percolates quickly through the sand, leaving behind many dying fish. Some Totonac Indians actually make a living by exploiting this easy fish harvest. In connection with observations of this kind, it is easy to understand that fish are also buried by sediments transported at the same time that they are themselves are. Forchhammer, Walther, Abel, and Deecke have already reported this finding, and Agassiz described the mass death of marine creatures on the coast of Florida, huge numbers of which were buried alive under the beach sand. This happens especially often in the swales behind beach walls from which the water can escape by filtration. At such places, I have often observed pronounced changes in temperature and salinity, which can be significant causes of chemical changes. Bube (30) told me about dead fish washed ashore along the coast near Seattle in such masses that they were exploited industrially to make glue and other products. According to Walther (170), Pakapy reported that in 1878, during a storm near Messina [Italy], seventeen hundred deep-sea fish were washed ashore.

An account such as this can never be complete, for most events of this nature take place without ever being observed and reported in a way that is of value to science.

Conclusion

This is the end of my observations. As I wrote this report, my main consideration was to stimulate interest in a method, and in this respect, my descriptions and pictures are only symbols of a methodology that our sciences of geology and paleontology can continue to use to great advantage, and in so doing, win new friends. We have scrutinized fossil remains, referring to recent examples to explain the peculiarities of their deposition.

This is not intended to be an exhaustive report. Completeness cannot and should not be the goal. Even in pictures I could not capture everything I saw—many impressive scenes remained unphotographed. For example, once, after a strenuous trip to the northern shore of Oyster Lake, at the point where Matagorda and Tres Palacios bays come together, I came upon a bare, salt-clay flat at the edge of the ocean, brightly lit by the setting sun. The norther had driven herds of cattle to this shore where they were caught fast in the moist clay and died of sickness or starvation. There they all lay, undisturbed, in their characteristic positions. Newly hatched sandpiper chicks from nearby nests ran peeping among the carcasses, and farther to the east, an osprey plunged earthward. It was a picture one does not soon forget. I remember another evening journey to the eastern shore of Lake Austin, south of the imposing Hawkins estate—now deserted and, according to the superstitions of the Negroes, haunted; its owners used to land their ships in the bay, long since silted in. The water in a creek entering from the east was low, as it was in the lake. The sticky clay soil was drying out, and a white line of salt efflorescence marked the meeting of land and water. Decaying cattle carcasses lay in the water, while on shore, some of those still alive wandered around dazed and drank the dangerous brackish water to quench their terrible thirst; others, fatally stricken, lay there dying—all because of a sudden change in the weather. Not far away, to the east, scattered herds were grazing and calves were frisking about in the sparse meadows.

This new field of study, even though it may at first seem disturbing and unaesthetic, is not without its fascination; I hope that this necessarily brief book, which clearly needs development, may be successful in promoting the method. Above all, I hope that soon new observations, so important for biologists and paleontologists, will be forthcoming from other quarters and will lead us to revisit and restudy the treasures of our museums and private collections, most of which we still do not understand completely. To this end, I have also tried to make accessible to the reader in comprehensible form the many reports from different sources relating to the subject at hand.

Figures

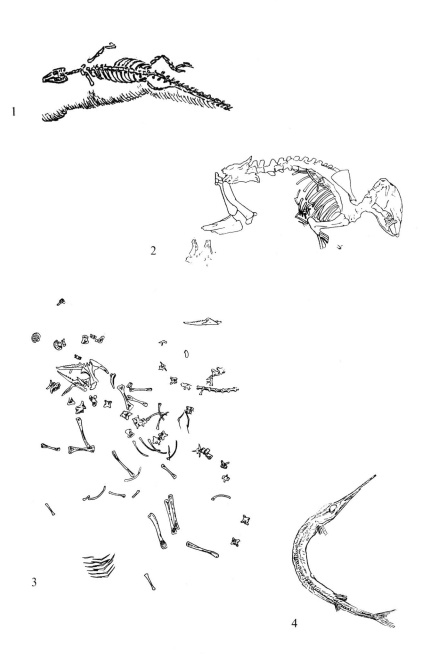

FIGURE 1.　Skeleton of *Homoeosaurus brevipes* H. von Meyer showing an impression tracing the outline of the body; drawing by Rothpletz.

FIGURE 2.　*Lagomys oeningensis* H. von Meyer *[Prolagus oeningensis]* from the Molasse at Oeningen.

FIGURE 3.　Completely disarticulated skeleton of *Homoeosaurus maximilianus* von Ammon from the Solnhofen lithographic limestone at Painter Forst.

FIGURE 4.　*Belonostomus tenuirostris* Ag.; juvenile from the lithographic limestone at Solnhofen.

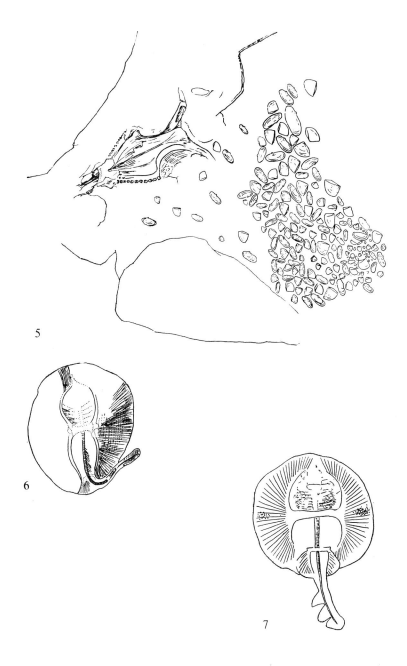

5

6

7

FIGURE 5.   Skull of a small nothosaurid from the Muschelkalk of the Nietleben syncline near Halle [East Germany] with a wedge of shelly debris in the flow shadow.

FIGURE 6.   *Urolophus crassicauda* Blainville from the Eocene of Monte Bolca, after Jaekel.

FIGURE 7.   *Platyrhina egertoni* de Zigno sp. from the Eocene of Monte Bolca, after Jaekel.

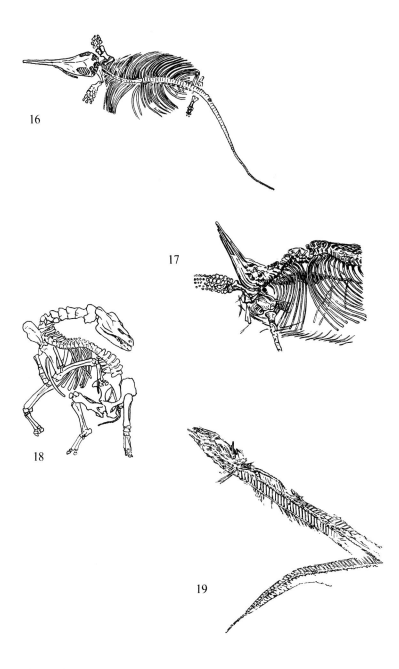

FIGURE 16.   *Ichthyosaurus tenuirostris* Owen from the English Liassic.

FIGURE 17.   Compression of the vertebral column in *Stenopterygius crassicostatus* Fraas, in the Senckenberg-Museum in Frankfurt, after Drevermann.

FIGURE 18.   Skeleton of *Palaeotherium magnum* Cuvier from the upper Eocene of Mormoiron, in France, after Abel.

FIGURE 19.   Kinked carcass of *Lepidopus glaronensis* Ag., ¼ natural size, after Wettstein.

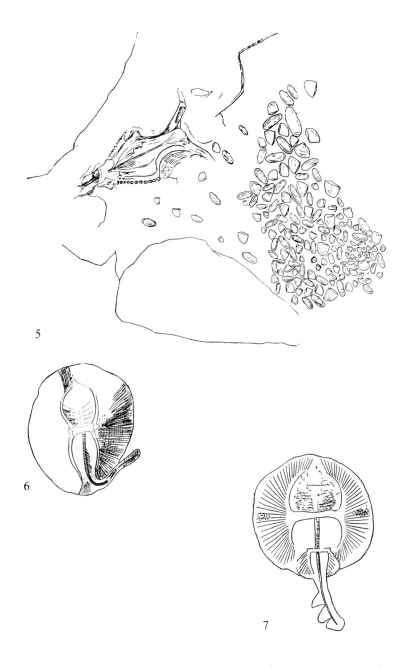

FIGURE 5. Skull of a small nothosaurid from the Muschelkalk of the Nietleben syncline near Halle [East Germany] with a wedge of shelly debris in the flow shadow.

FIGURE 6. *Urolophus crassicauda* Blainville from the Eocene of Monte Bolca, after Jaekel.

FIGURE 7. *Platyrhina egertoni* de Zigno sp. from the Eocene of Monte Bolca, after Jaekel.

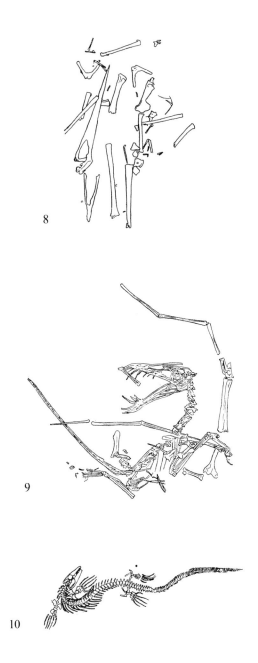

FIGURE 8.   Skeleton of a *Pteranodon* from the Niobrara Group [Niobrara Chalk] in Kansas, after Wiman.

FIGURE 9.   Skeleton of *Dorygnathus banthensis* W., after Wiman, showing the influence of flowing water.

FIGURE 10.   Skeleton of *Platecarpus coryphaeus* Marsh from the Niobrara Group [Niobrara Chalk] in Kansas, after Wiman.

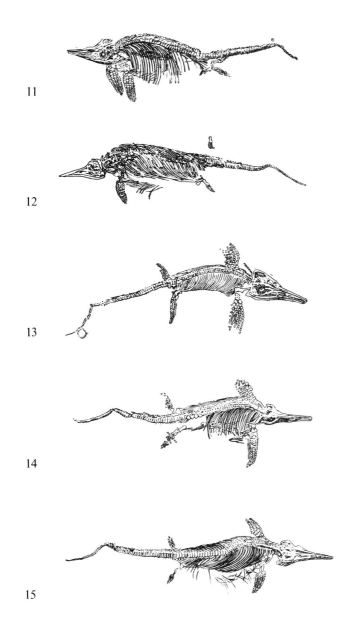

FIGURE 11. *Stenopterygius* sp. from the Posidonia Shale at Holzmaden, after Huene.

FIGURE 12. *Stenopterygius hauffianus* von H., Posidonia Shale at Holzmaden.

FIGURE 13. *Leptopterygius disinteger* von Huene with a curl in the caudal vertebral column. Posidonia Shale at Holzmaden.

FIGURE 14. *Stenopterygius crassicostatus* E. Fraas from the Posidonia Shale at Holzmaden, after Huene.

FIGURE 15. *Stenopterygius zetlandicus* Seeley from the Posidonia Shale at Ohmden near Holzmaden, after Huene.

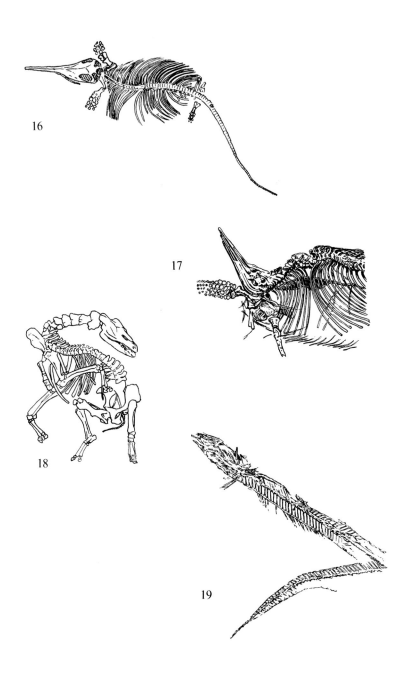

FIGURE 16.   *Ichthyosaurus tenuirostris* Owen from the English Liassic.

FIGURE 17.   Compression of the vertebral column in *Stenopterygius crassicostatus* Fraas, in the Senckenberg-Museum in Frankfurt, after Drevermann.

FIGURE 18.   Skeleton of *Palaeotherium magnum* Cuvier from the upper Eocene of Mormoiron, in France, after Abel.

FIGURE 19.   Kinked carcass of *Lepidopus glaronensis* Ag., ¼ natural size, after Wettstein.

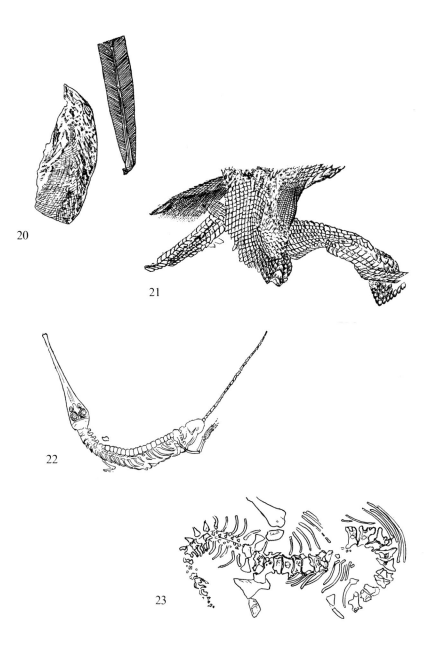

20

21

22

23

FIGURE 20. *Taeniopteris* leaf and *Palaeoniscum* buried axially parallel in the Mansfeld Kupferschiefer, from the Wolfschacht mine near Eisleben; ½ natural size.

FIGURE 21. Shred of scaly skin from an *Acrolepis asper* Ag. after a specimen in the Bergschule, at Eisleben; the counterpart is in the collection of the University of Greifswald; ½ natural size.

FIGURE 22. Curved skeleton of *Teleosaurus chapmani* Owen.

FIGURE 23. Vertebral column of *Procolophon* showing double curve; after Seeley.

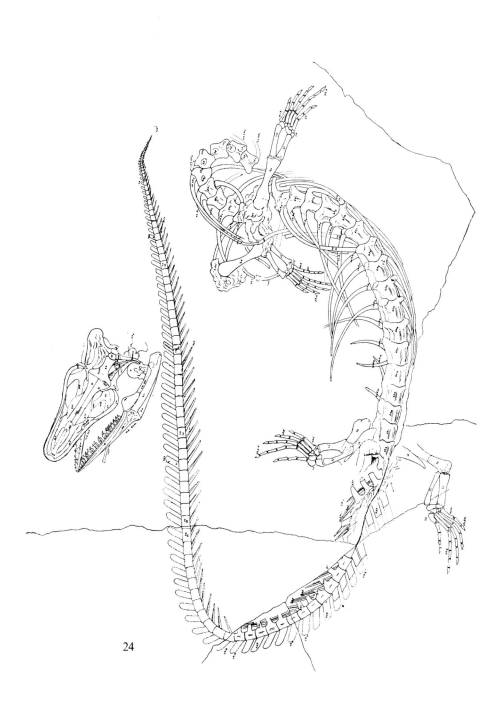

24

FIGURE 24. *Opetiosaurus bucchichi* Kornhuber from the Lower Cretaceous of Lesina, in Dalmatia, ⅓ natural size.

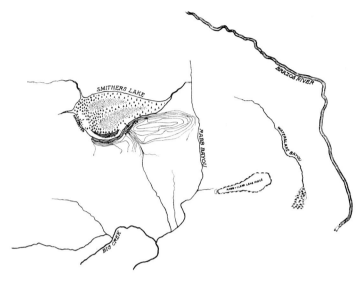

25

FIGURE 25.   Map showing Smithers Lake, in Texas.

26

27

28

FIGURE 26. Skeleton of *Branchiosaurus amblystomus* Cr. [*Apateon* sp.] from the Rotliegend at Friedrichroda, after Langenhahn.

FIGURE 27. Skeleton of a *Protriton* [*Apateon* sp.] from the Rotliegend at Friedrichroda, after Langenhahn.

FIGURE 28. Group of six individuals of *Pantelosaurus saxonicus* von Huene [*Haptodus baylei*] from the Lower Permian of the Queen Carola Mine, near Döhlen, in the Plauenscher Grund. 1/12 natural size.

Plates

A

B

C

Plate 1

FIGURE A. Fresh cow carcass a few hours after death on the prairie near Lockwood, Texas. February 1925.

FIGURE B. Collapsed cow carcass on the prairie near Beasley, Fort Bend County; the right foreleg has been moved away from the rest of the carcass. February 1925.

FIGURE C. Mummified forelegs of a cow, from the region around Fulshear, north of the Brazos River, Fort Bend County. February 1925.

A

B

C

D

Plate 2

FIGURE A.   Whale carcass washed up parallel to the shore, partially embedded between the pectoral girdle and the pelvis; coast of the Gulf of Mexico between the mouths of the Sabine and Calcasieu rivers. DUFF. January 1925.

FIGURE B.   Bird carcass in driftwood; carcass assemblage at Smithers Lake. February 1925.

FIGURE C.   Bird carcass with ganoid and teleost fishes in the strandline of the lake; carcass assemblage at Smithers Lake. The plumage has become quite inconspicuous. February 1925.

FIGURE D.   Cow that died after being stuck in a small swampy spot for three days, with characteristic evidence of vulture feeding in the pelvic region. Prairie near Beasely. February 1925.

A

B

C

Plate 3

FIGURE A. Cow stuck in the mud of the salt marsh on the Intracoastal Canal at Sargent, Matagorda County. Died December 20, 1924. October 10, 1925.

FIGURE B. Cow carcass with stomach contents visible, flanks damaged by feeding wolves. Prairie near Bay City, Matagorda County. March 21, 1925.

FIGURE C. Cow that died in quicksand; only the horns and nasals can be seen. Gulf Coast east of Sargent, on the border between Matagorda and Brazoria Counties. May 10, 1925.

A

B

C

Plate 4

FIGURE A. Partial view of a quicksand carcass assemblage, exposed by the wind; coast of the Gulf of Mexico east of Sargent between the Bernard River and Matagorda Bay. August 8, 1925.

FIGURE B. Quicksand carcass near Sargent. May 10, 1925.

FIGURE C. Quicksand carcass assemblage near Sargent; cattle skull showing calcareous crust in the eye sockets, lying on its side, with sand crab tracks. August 8, 1925.

A

B

C

D

E

Plate 5

FIGURE A.   Cow skull from the quicksand carcass assemblage east of Sargent, showing stepwise growth at the base of the horns. August 8, 1925.

FIGURES B, C, D.   Pelvises in dorsal positions, in some places showing mummified remains of tendons and muscles; exposed by wind. Quicksand carcass assemblage east of Sargent. August 8, 1925.

FIGURE E.   Partially mummified carcass from a part of the quicksand bed that dried out quickly. East of Sargent. August 8, 1925.

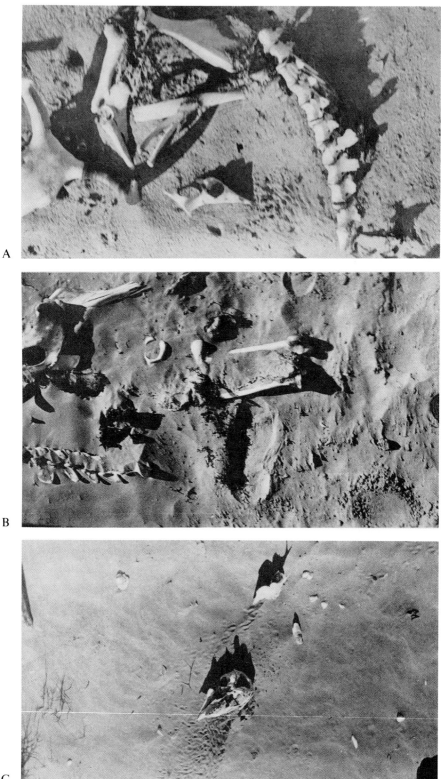

A

B

C

Plate 6

FIGURE A. Cow carcass exposed by the wind at the quicksand carcass assemblage east of Sargent; next to it, the spindle from a *Fusulina.* August 8, 1925.

FIGURE B. Badly buckled carcass from the quicksand carcass assemblage east of Sargent. August 8, 1925.

FIGURE C. Removal of the jawbone of a calf skull by sand crabs, whose tracks are visible; quicksand carcass assemblage east of Sargent. August 8, 1925.

A

B

C

D

Plate 7

FIGURE A. Pike carcass in flattened grass; flooding of the Elbe in the autumn of 1926, near Dessau. E. Voigt.

FIGURE B. Pike carcass whose backbone was torn out by feeding crows; autumn flooding of the Elbe, 1926. E. Voigt.

FIGURE C. Erosional cavity and desiccated mud flat with ganoid fish; old rice canal near Bay City. May 7, 1925.

FIGURE D. Erosional cavity with needle gar, lateral view; near Bay City, Matagorda County. May 7, 1925.

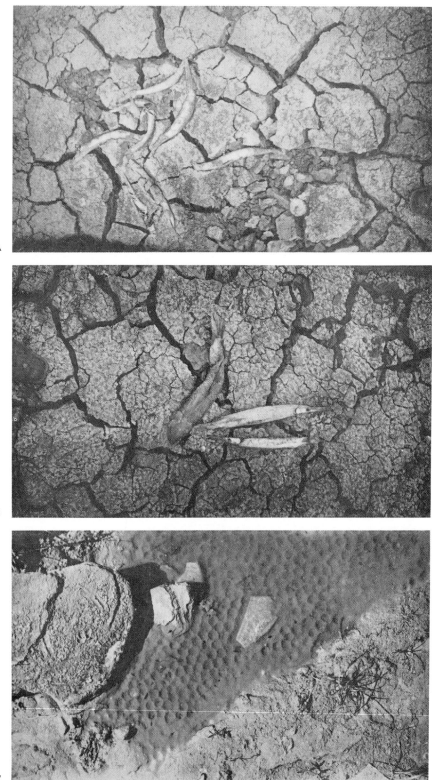

A

B

C

Plate 8

FIGURE A.   Top view of needle gar in erosional cavity, near Bay City. May 7, 1925.

FIGURE B.   Dried-up erosional cavity with two needle gars and a bowfin; the ground looks marbled because of dried bird tracks, which predate the mud cracks; area around Bay City. May 7, 1925.

FIGURE C.   Depressions made by tadpoles in the drying mud of the banks of the Brazos River between Rosenberg and Richmond. April 18, 1925.

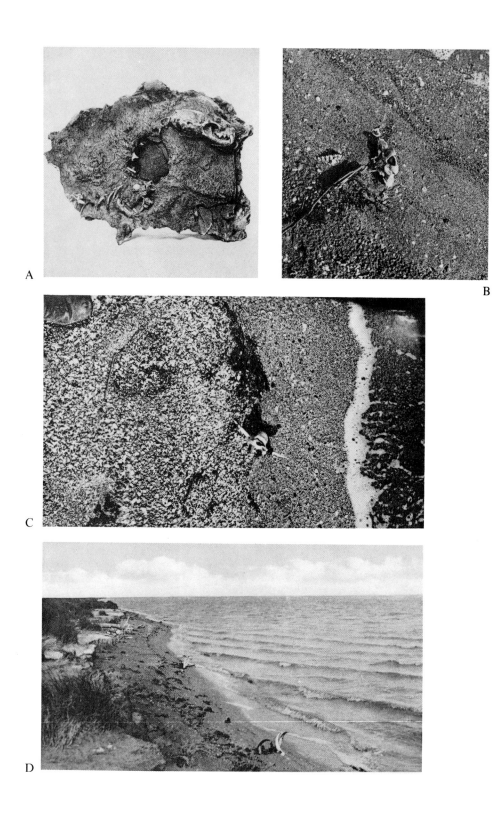

A

B

C

D

Plate 9

FIGURE A. Carcass of a weasel from a Kupferschiefer mine shaft; in the middle is a drip hole from which the cadaver was sintered with lime; Eisleben. Bessler collection.

FIGURE B. Cow skull embedded in the margin of the beach on the north shore of the eastern end of Matagorda Bay; Baer Ranch. September 3, 1925.

FIGURE C. Cow vertebra on the margin of the beach at the east end of Matagorda Bay; Baer Ranch. September 3, 1925.

FIGURE D. Beach on the north shore of Matagorda Bay near its eastern end; erosional coastline with buried cattle carcasses. September 3, 1925.

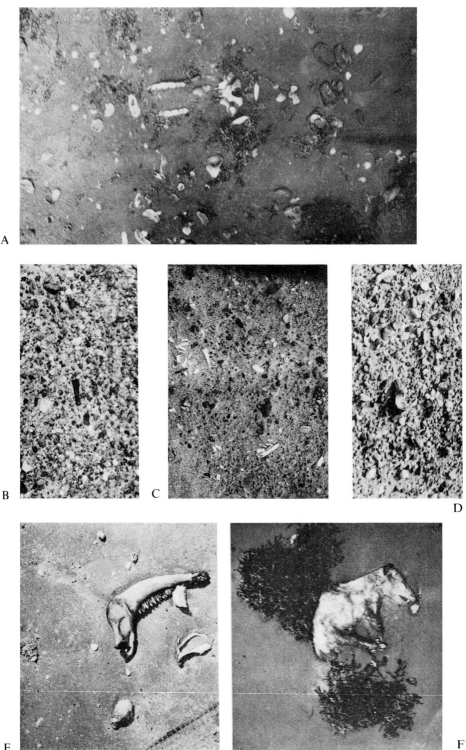

A

B

C

D

E

F

Plate 10

FIGURE A.   Cow skull embedded in the beach of the Gulf of Mexico, surrounded by sargasso and oysters; near the mouth of the Bernard River, Brazoria County. October 10, 1925.

FIGURE B.   Pleistocene bone in the bed of the Brazos above Richmond. February 20, 1925.

FIGURE C.   Disintegrating fossil ivory on a pebble bank in the riverbed of the Brazos above Richmond. April 9, 1925.

FIGURE D.   Pleistocene bones in a gravel bank in the bed of the Brazos near Richmond. April 20, 1925.

FIGURE E.   Lower jaw of a cow with shells of the Virginia oyster; coast of the Gulf of Mexico near Cameron, Louisiana. January 1925.

FIGURE F.   Opossum carcass in a passive position, with seaweed; beach of the Gulf of Mexico near the mouth of the Bernard River. May 10, 1925.

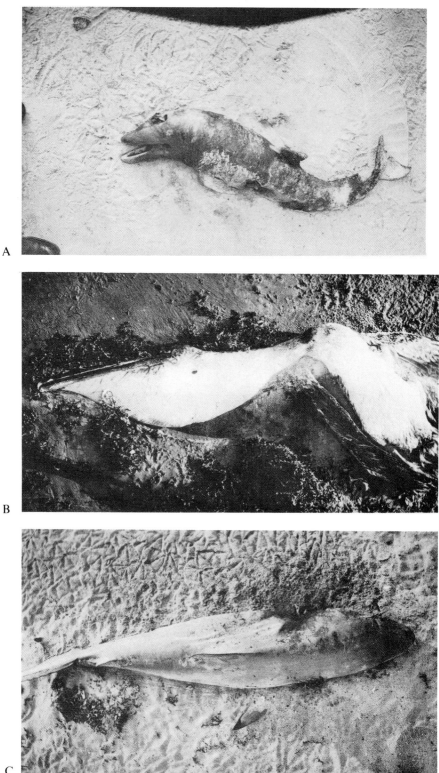

A

B

C

Plate 11

FIGURE A.   Porpoise fed upon by scavenging gulls, with tracks and drag marks in the sand; beach of the open gulf near Sargent. February 25, 1925.

FIGURE B.   Beached pelican carcass with sand-filled throat pouch and an indentation in the sand that follows the outline of the body; Gulf Coast between Sargent and Mitchell's Cut. May 5, 1925.

FIGURE C.   Shark carcass on the Gulf of Mexico; left fin was torn off by scavenging gulls. May 7, 1925.

A

B

C

D

Plate 12

Fɪɢᴜʀᴇ A. Carcass of a large alligator gar, displaced by feeding vultures; carcass assemblage at Smithers Lake. February 1925.

FɪɢᴜʀᴇB. Mummified carcass of a soft-shelled turtle showing the neck bent sharply backward; Smithers Lake. March 15, 1925.

FɪɢᴜʀᴇC. Skeleton of a wild duck with hooked backbone; carcass assemblage at Smithers Lake. March 15, 1925.

FɪɢᴜʀᴇD. Shark carcass on the Gulf Coast near Freeport, Brazoria County; accumulation of shells dense on one side, sparse on the other. April 13, 1925.

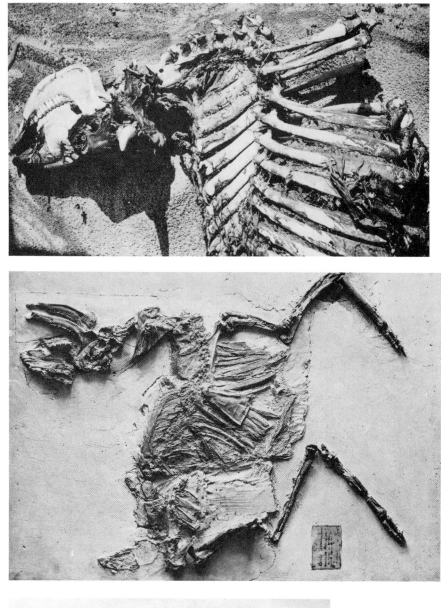

A

B

C

Plate 13

Fᴵɢᵁᴿᴇ A.   Cow carcass with neck bent sharply backward, on a sandbank of the Brazos River north of Brazoria. March 21, 1925.

Fɪɢᴜʀᴇ B.   *Cervus (Diceroceras) furcatus* O. Fr. [*Euprox furcatus* Hensel] from the Miocene of the Steinheim Basin; size: 80 cm × 107 cm, original in the natural history collection at Stuttgart [Staatliches Museum für Naturkunde].

Fɪɢᴜʀᴇ C.   Camel carcass with neck bent backward, in the stone desert between Aswan and the oasis of Curcur. Dienst, March 1909.

A

B

C

Plate 14

FigURE A. Ganoid fish carcass at the boundary line between an accumulation of broken shells and fine sand; coast of the Gulf of Mexico near the mouth of the Calcasieu River, in Louisiana. January 1925.

FigURE B. Cow carcass on the beach of the Gulf of Mexico, partially embedded, skeletonized by sand crabs; the skin has dried up and looks like parchment. 1925.

FigURE C. Stingray carcass on the inner side of a beach ridge; Mitchell's Cut, at the eastern outlet of Matagorda Bay. April 19, 1925.

A

B

C

D

E

Plate 15

FIGURE A.   Pelican carcass with alluvial debris trail; coast of the Gulf of Mexico near Cameron, Louisiana, west of the mouth of the Calcasieu River. January 25, 1925.

FIGURE B.   *Aspidorhynchus acutirostris* Blainville from the Solnhofen lithographic limestone; badly broken and bent into a hook shape, with backbone torn out and running from the tail to below the head. Collection of the Geological Institute at Halle [East Germany].

FIGURE C.   Tern carcass with outspread wings, surrounded by sargasso; beach of the Gulf of Mexico. May 17, 1925.

FIGURE D.   *Aspidorhynchus acutirostris* Blainville from the Solnhofen lithographic limestone; bent into a hook shape with displacement of the posterior parts of the backbone. Collection of the Geological Institute at Halle.

FIGURE E.   Buckled carcass of *Pygopterus* from the Kupferschiefer at Eisleben. Pangert collection, Eisleben.

A

B

C

Plate 16

FIGURE A. Horse in dorsal position with ribs spread apart, in an oxbow of the Brazos, near Fulshear, Texas. April 19, 1925.

FIGURE B. Cow carcass hanging over the edge of a beach cliff that is being washed away; Baer Ranch, eastern end of Matagorda Bay. September 3, 1925.

FIGURE C. Coyote carcass, distorted by desiccation; near Fulshear, Texas. April 19, 1925.

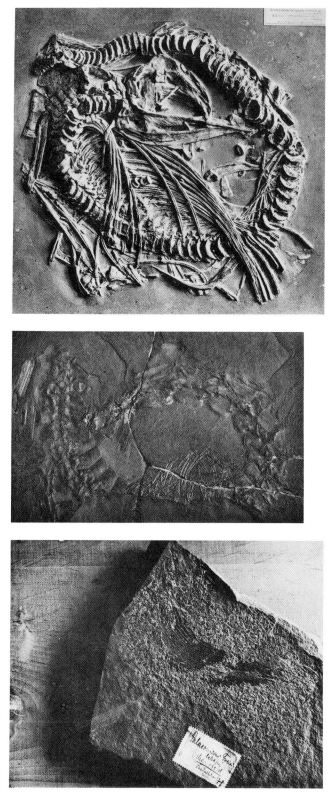

A

B

C

Plate 17

A

B

C

D

Plate 18

FIGURE A.   View of the southern shore of Indian Bay on Prien Lake, south of Lake Charles, Louisiana, at the onset of a norther on December 19, 1924. The surface water vaporized and was blown to the southern shore, where it formed ice.

FIGURE B.   Weathered tree damaged by northers—live oak on the Lockwood prairie; old live oaks on the prairie are always better developed on the side facing south; the side facing north is thinned out or has died off due to ice buildup. February 1925.

FIGURE C.   West shore of Smithers Lake, showing tree stumps from the forest that has died out. March 15, 1925.

FIGURE D.   Same as figure C.

A

B

C

D

Plate 19

Figure A.   Stump horizon on the western part of the south shore of Smithers Lake, looking toward the west shore. March 15, 1925.

Figure B.   View of the west shore of Smithers Lake, showing remains of the destroyed forest. March 1925.

Figure C.   View from the west shore of Smithers Lake, looking north. March 1925.

Figure D.   Carcasses of two large alligator gars at the edge of the carcass assemblage at Smithers Lake, as the water is going down. February 1925.

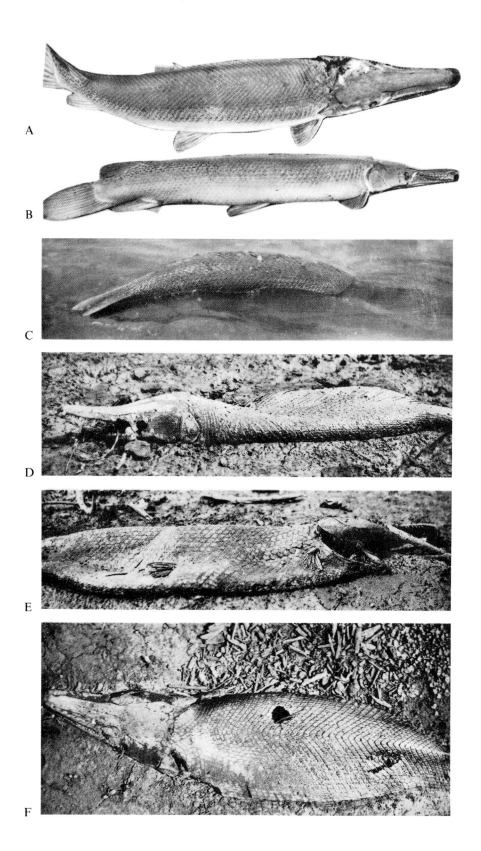

A

B

C

D

E

F

Plate 20

FIGURE A. Female alligator gar caught and shot near Tres Palacios; top view.

FIGURE B. Same as figure A, side view.

FIGURE C. Side view of alligator gar near Mitchell's Cut, near the eastern outlet of Matagorda Bay. May 17, 1925.

FIGURE D. Collapsed carcass of an alligator gar; it decomposed in water and later dried out; the underside of the head is shown; the tail lies sideways, hence the fold. March 15, 1925.

FIGURE E. Side view of a similar carcass. March 15, 1925.

FIGURE F. Accumulation of plant remains in the flow shadow of a gar carcass in ventral position. March 1925.

A

B

C

Plate 21

Figure A.   Alligator gar lying perpendicular to the shore. It dried out fairly soon after death; the mouth touches a band of woody debris; the lower jaw and one fin were carried off by vultures; Smithers Lake. March 15, 1925.

Figure B.   Alligator gar carcass perpendicular to the bank, embedded in a band of woody debris; Smithers Lake. March 15, 1925.

Figure C.   Juvenile alligator gar in dorsal position; throat scales disintegrating. March 15, 1925.

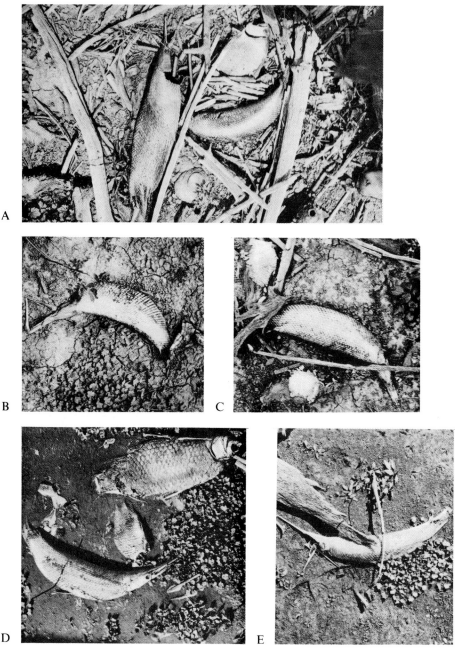

A

B

C

D

E

Plate 22

FIGURE A. Side view of needle gar in carcass assemblage at Smithers Lake; curved, ventral position. March 1925.

FIGURE B. Sharply curved carcass of a needle gar, showing disarticulated lower jaw. March 1925.

FIGURE C. Curved needle gar from the carcass assemblage at Smithers Lake. March 15, 1925.

FIGURE D. Needle gar and teleost fish at Smithers Lake; water level going down. February 1925.

FIGURE E. Characteristic decay of the skull of a needle gar; decay took place underwater. February 1925.

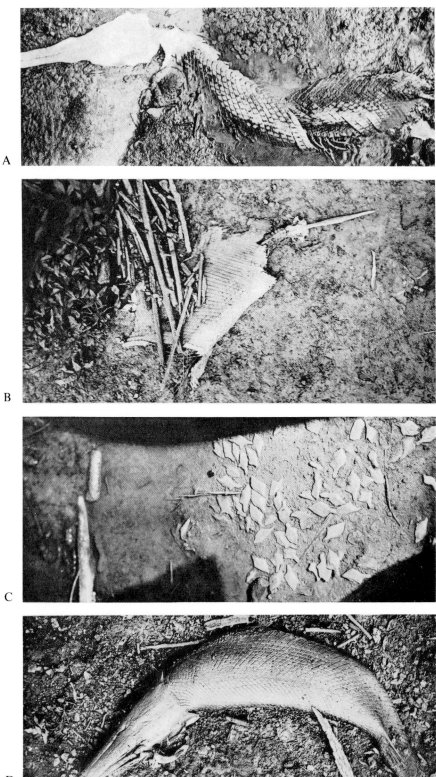

A

B

C

D

Plate 23

FIGURE A. Alligator gar carcass curved into an S shape; on the convex side, part of the backbone is visible; carcass assemblage at Smithers Lake. March 15, 1925.

FIGURE B. Advanced state of decay of alligator gar carcass that was underwater for a long time; the parasphenoid lies perpendicular to the strandline; the carcass itself bends around to line up with the strandline; Smithers Lake. March 15, 1925.

FIGURE C. Complete disintegration of the scaly skin of an alligator gar, in the carcass assemblage at Smithers Lake. March 21, 1925.

FIGURE D. Curvature of the carcass of an alligator gar; the carcass dried out soon after death and was mummified. March 15, 1925.

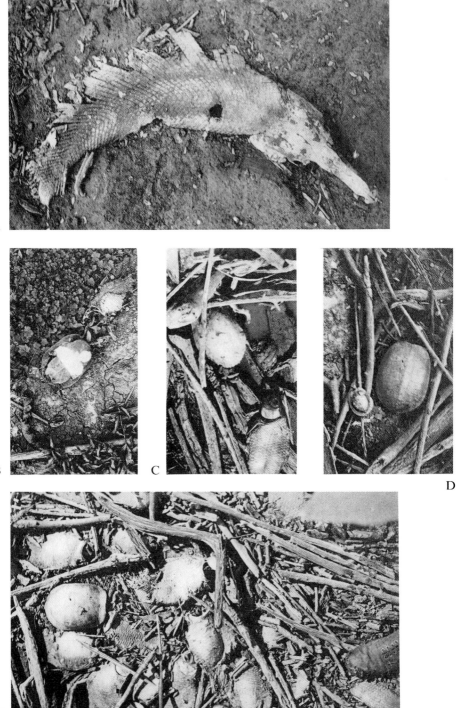

A

B

C

D

E

Plate 24

FIGURE A. Characteristic disintegration of an alligator gar in a curved ventral position; the creature lay in water for a long time; the mouth points toward the lake; a band of plant debris begins at the tail. March 1925.

FIGURE B. Turtle in a dorsal position, showing the skull extended and the shell flaking, lying axially parallel to a teleost fish. March 15, 1925.

FIGURE C. Turtle in a band of woody debris. February 1925.

FIGURE D. Turtle from the carcass assemblage at Smithers Lake. March 19, 1925.

FIGURE E. Turtle in a mosaic of fish at Smithers Lake. February 1925.

A

B

C

D

Plate 25

FIGURE A.   Mud turtle, dried out quickly and mummified. February 1925.

FIGURE B.   Carcass of a frozen mud turtle, in the carcass assemblage at Smithers Lake. February 1925.

FIGURE C.   Curved swath of debris washed ashore, carcass assemblage at Smithers Lake. February 1925.

FIGURE D.   Teleost fish mosaic, on the margin of the shore at Smithers Lake. February 1925.

A

B

C

Plate 26

FIGURE A.   Angular strandline, Smithers Lake; in the left background are the two large gars shown in plate 29, figure C. February 1925.

FIGURE B.   Swath of driftwood and carcasses on the south shore of Smithers Lake. February 1925.

FIGURE C.   Large alligator washed up perpendicular to the shore; the snout points toward the water; a needle gar is behind the tail; carcass assemblage at Smithers Lake. Early February 1925.

A

B

C

Plate 27

FIGURE A. Alligator carcass rotated on its longitudinal axis by the current; Smithers Lake. February 1925.

FIGURE B. Alligator carcass forced into an angular, dorsal position; Smithers Lake. February 1925.

FIGURE C. Curved alligator carcass in a dorsal position; carcass assemblage, Smithers Lake. February 1925.

A

B

C

Plate 28

FIGURE A. Alligator carcass lying perpendicular to the shore; the posterior end is mummified and bends following the shoreline; the anterior half of the body and the forelegs are more extensively macerated; the right legs are more extended than the left; Smithers Lake. March 15, 1925.

FIGURE B. Alligator washed up perpendicular to the shore, the left half of the body somewhat sunk in mud; in February, it was still lying in shallow water. February 1925.

FIGURE C. Same animal a month and a half later; the right legs are extended, the left flexed; as the water went down, it left a buildup of debris partially around the front legs. March 21, 1925.

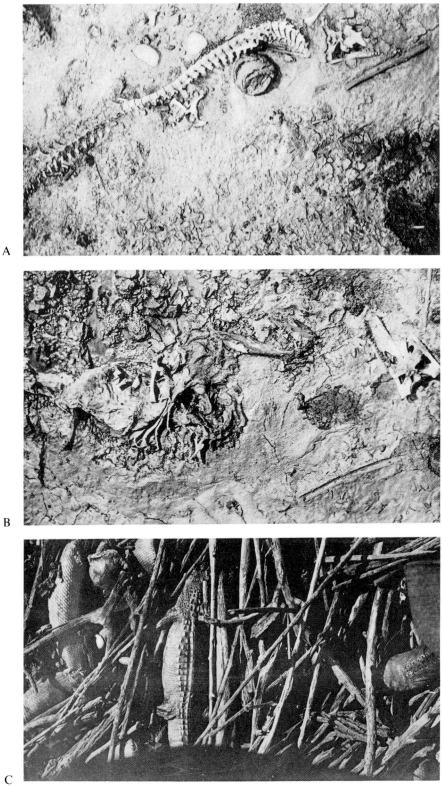

A

B

C

Plate 29

FIGURE A. Completely skeletonized carcass of a large alligator in a radial position; the tail lies sideways, the skull dorsal, and the neck vertebrae are correspondingly twisted; the animal lay in water to the very end; carcass assemblage at Smithers Lake. March 15, 1925.

FIGURE B. Alligator carcass; the front part lay in water a long time but finally dried out; the head was skeletonized and displaced; the posterior end is more mummified; the spine is beginning to curve. March 15, 1925.

FIGURE C. Parallel embedding of woody debris, fish, and alligator carcasses on the edge of the shore at Smithers Lake. February 25, 1925.

A

B

C

D

Plate 30

FIGURE A.   Alligator in a dorsal position lying tangential to driftwood; carcass assemblage at Smithers Lake. February 12, 1925.

FIGURE B.   Alligator and needle gar lying perpendicular to each other, embedded in the driftwood line at Smithers Lake. February 19, 1925.

FIGURE C.   The same carcass as in figure B, a month later. March 13, 1925.

FIGURE D.   Young alligator and a needle gar; notice their depositional positions relative to the driftwood. February 1925.

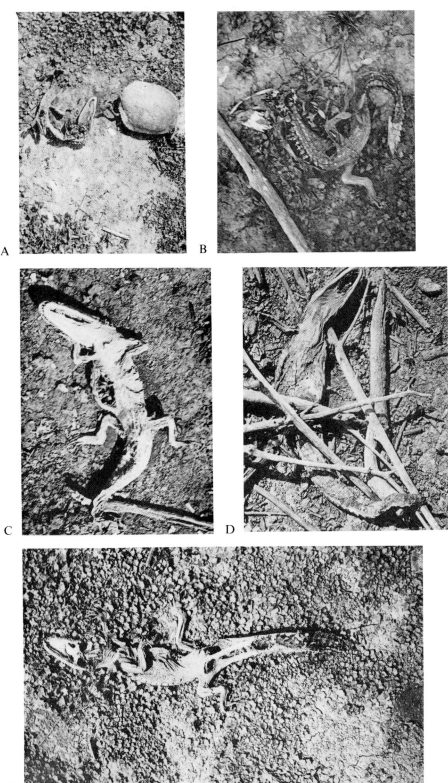

Plate 31

FIGURE A. Mummified alligator carcass showing circular curvature like the protorosaurian from the Kupferschiefer, with remains of a turtle; shore side of the carcass assemblage at Smithers Lake. March 15, 1925.

FIGURE B. Mummified carcass of a young alligator, showing S-shaped curvature, from the shore zone of the carcass assemblage at Smithers Lake. March 19, 1925.

FIGURE C. Mummified alligator carcass; carcass assemblage at Smithers Lake. March 15, 1925.

FIGURE D. Mummified alligator carcass; carcass assemblage at Smithers Lake. March 15, 1925.

FIGURE E. Mummified alligator carcass in extended but asymmetrical, dorsal position. March 15, 1925.

A

B

Plate 32

FIGURE A. Slab with *Neusticosaurus pygmaeus* Fraas (nine individuals); slab from the Lettenkohle of Hoheneck, near Eggolsheim; original slab in the natural history collection at Stuttgart [Staatliches Museum für Naturkunde].

FIGURE B. Slab with a group of *Neusticosaurus pusillus* Fraas, from the Lettenkohle of Eggolsheim; original slab in the natural history collection at Stuttgart.

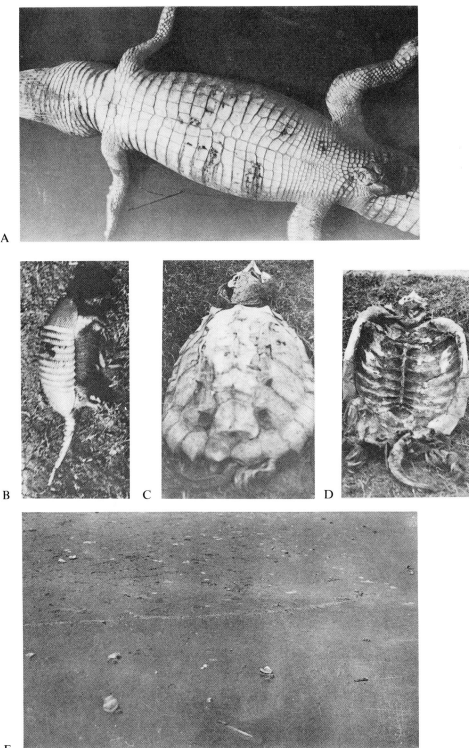

A

B   C   D

E

Plate 33

FIGURE A.   Alligator carcass inflated by gas, dorsal position; carcass assemblage at Smithers Lake. Early February, 1925.

FIGURE B.   Armadillo carcass (*Dasypus novemcinctus*) inflated by gases produced by decay; wooded area east of Bay City, Texas. 1925.

FIGURES C, D.   *Macrochelys temmincki* Holbrook; salt marsh southeast of Lake Charles, Louisiana. 1925.

FIGURE E.   Mud flat at the east end of Matagorda Bay, during rising water level; the white spots in the background are bird droppings; at the edge of the water is a whitish border composed of remains of fish and crab; there are no bird droppings in the foreground. May 17, 1925.

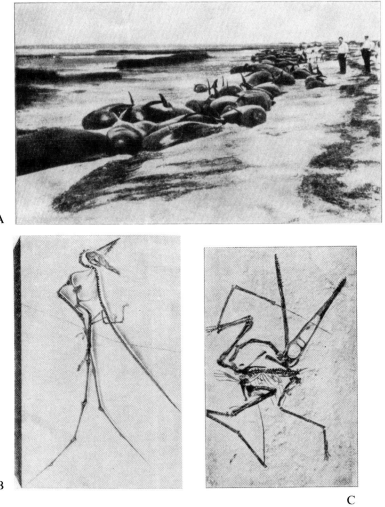

A

B

C

Plate 34

FIGURE A. School of seventy-five black whales that swam ashore on the coast of Massachusetts (*Berliner Illustrierte Zeitung*, 34th year, no. 35).

FIGURE B. *Rhamphorhynchus longicaudatus* Münst. showing the skull turned to the right, wings becoming detached; from Eichstätt (after Wagner).

FIGURE C. Sharp backward bending of the neck in *Pterodactylus longirostris*.

Plate 35

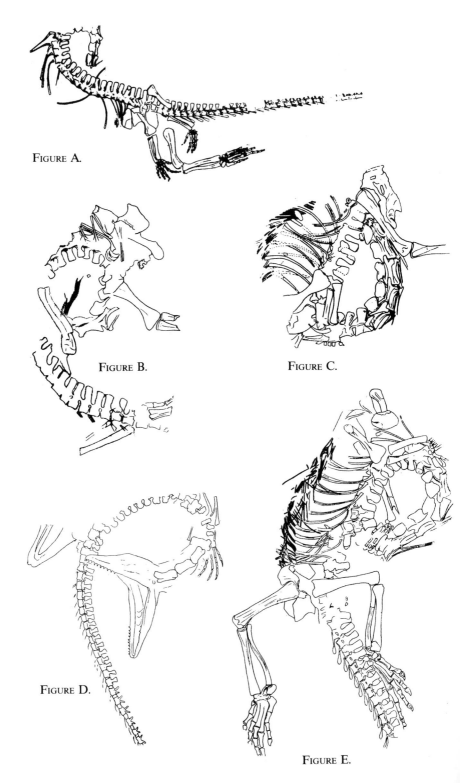

FIGURE A.

FIGURE B.

FIGURE C.

FIGURE D.

FIGURE E.

Plate 36

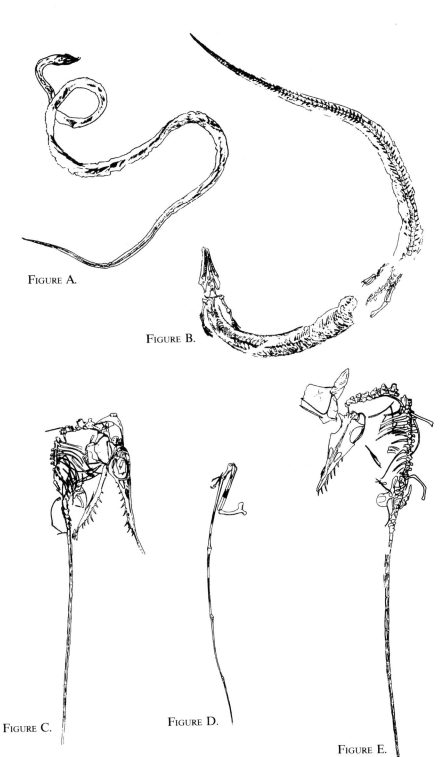

FIGURE A.

FIGURE B.

FIGURE C.

FIGURE D.

FIGURE E.

Plate 37

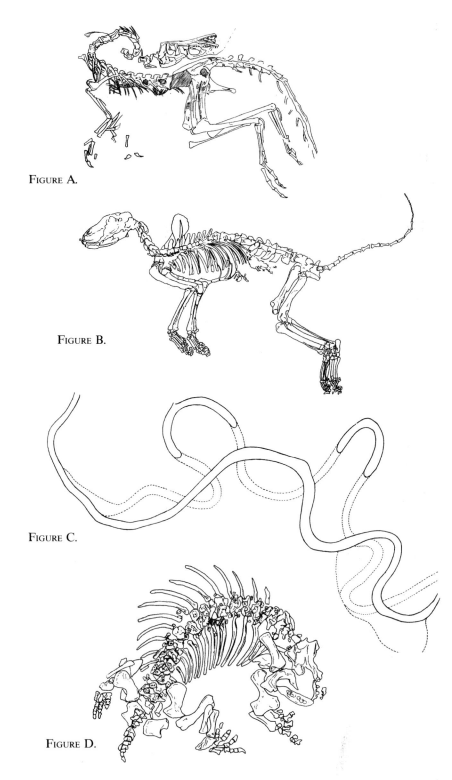

FIGURE A.

FIGURE B.

FIGURE C.

FIGURE D.

References

Weigelt's numerically coded system of referencing has been retained in this translation to permit cross-referencing to the German original. A number of entries do not appear in alphabetical or chronological sequence in the bibliography to Weigelt's book. I have not attempted to rectify this problem because it would necessitate confusing changes in the numbering. Many of the references provided by Weigelt lack important data for literature search or are incorrect. The bibliography thus has been carefully revised as far as possible.

Weigelt cited specific passages of text or figures in some cases; these are marked by an asterisk and are placed in brackets at the end of the relevant bibliographic entries.

H.-D. S.

1. Abel, O. 1908. Die Anpassungsform der Wirbeltiere an das Meeresleben. Vortr. Ver. Verbr. Naturw. Kenntnisse Wien, 48: 395–422.
2. Abel, O. 1913. Die eocänen Sirenen der Mittelmeerregion. Erster Teil. Der Schädel von *Eotherium aegyptiacum*. Palaeontographica, 59: 289–360.
3. Abel, O. 1919. Die Stämme der Wirbeltiere. XVIII + 914 pp. Walter de Gruyter & Co.; Berlin and Leipzig.
4. Abel, O. 1922. Lebensbilder aus der Tierwelt der Vorzeit. VIII + 643 pp. Gustav Fischer; Jena.
5. Abel, O. 1924. Die Rekonstruktion von *Palaeotherium magnum* Cuvier. Paläont. Z., 6: 14–24.
6. Abel, O. 1926. Diskussionsbemerkungen [concerning Weigelt's lecture]. Paläont. Z., 8: 328.
7. Abel, O. 1926. Amerikafahrt. Eindrücke, Beobachtungen und Studien eines Naturforschers auf einer Reise nach Nordamerika und Westindien. + 462 pp. Gustav Fischer; Jena.
8. Andreae, A. 1893. Vorläufige Mitteilung über die Ganoiden (*Lepidosteus* und *Amia*) des Mainzer Beckens. Verh. Naturhist. Med. Ver. Heidelberg, N.F., 5: 7–15.

9. Andrée, K. 1920. Geologie des Meeresbodens. Vol. II. xx + 689 pp. Gebrüder Borntraeger; Berlin.

10. Agassiz, L. 1833–43. Recherches sur les poissons fossiles. 5 vols. Imprimerie de Petitpierre; Neuchatel.

11. Ammon, L. v. 1885. Über *Homoeosaurus Maximiliani*. Abh. Kgl. Bayer. Akad. Wiss., II. Cl., 15: 497–528.

12. Ammon, L. v. 1884. Ueber das in der Sammlung des Regensburger naturwissenschaftlichen Vereines aufbewahrte Skelett einer langschwänzigen Flugeidechse (*Rhamphorhynchus longicaudatus*). Correspondenzbl. Naturwiss. Ver. Regensburg, 38: 129–67.

13. Ammon, L. v. 1889. Die permischen Amphibien der Rheinpfalz. 119 pp. F. Straub; Munich.

14./15. Ammon, L. v. 1907. Über jurassische Krokodile aus Bayern. Geogn. Jh., 18: 55–71.

16. Andrews, C. W. 1896. On the extinct birds of the Chatam islands. Parts I and II. Novit. zool., Lond., 3: 73–84, 260–71.

17. Antipa, G. 1911. Das Überschwemmungsgebiet der unteren Donau. Anuarul Instut. Geol. al Roman., 4(2) (1910): 225–496.

18. Assmann, P. 1906. Über *Aspidorhynchus*. Arch. Biontol., 1: 49–79.

19. Barnes, B. Eine eozäne Wirbeltier-Fauna aus der Braunkohle des Geiseltales. Jb. Hall. Verb. Erforsch. Mitteldeutsch. Bodenschätze, 6: 5–24.

20. Beebe, W. C. 1924. Galapagos: World's End. xix + 443 pp. G. P. Putnam's; New York. [German translation: Galapagos. Das Ende der Welt. Leipzig; 1926.]

21. Berendt, G. 1869. Geologie des Kurischen Haffes und seiner Umgebung. Königsberg.

22. Berz, K. 1926. Über die Natur und Bildungsweise der marinen Eisensilikate, insbesondere der chamositischen Substanzen. Fortschr. Geol. Paläont., 11: viii + 158 pp.

23. Besser, H. 1917. Natur- und Jagdstudien in Deutsch-Ostafrika.

24. Borissiak, A. 1914. [Mammifères fossiles de Sebastopol. I.] Bull. Com. Géol. St. Petersbourg, n.s., 87: xii + 154 pp. (In Russian with French summary.)
    Borissiak, A. 1915. [Mammifères fossiles de Sebastopol. II.] Bull. Com. Géol. St. Petersbourg, n.s., 137: 1–24 (Russian text), 25–47 (French text).

25. Braun, G. n.d. Leiter der Greifswalder Lappland-Expedition: 30000 hungernde Renntiere. Berliner Illustrierte Zeitung, no. 49: 1635.

26. Broili, F. 1925. Beobachtungen an der Gattung *Homoeosaurus* H. v. Meyer. Sitz.-Ber. Bayer. Akad. Wiss., Math.-Naturw. Abt., 1925: 81–121.

27. Broili, F. 1904. Permische Stegocephalen und Reptilien aus Texas. Palaeontographica, 51: 1–49.

28. Broili, F., and E. Fischer. 1917. *Trachelosaurus Fischeri* nov. gen. nov. sp. Ein neuer Saurier aus dem Buntsandstein von Bernburg. Jb. Preuss. Geol. Landesanst., 37: 359–414.

29. Broili, F. 1925. Ein *Pterodactylus* mit Resten der Flughaut. Sitz.-Ber. Bayer. Akad. Wiss., Math.-Naturw. Abt., 1925: 23–34.

29a. Broili, F. 1926. Ein neuer Fund von *Pleurosaurus* aus dem Malm Frankens. Abh. Bayer. Akad. Wiss., 30 (8): 1–48.

30. Bube, K. Oral communication.

31. Büchler, F. 1920–21. Beitrag zur Kenntnis der grünen und roten Letten. Doctoral dissertation, Universität Würzburg.

32. Credner, H. 1888. Die Stegocephalen und Saurier aus dem Rothliegenden des Plauen'schen Grundes bei Dresden. Siebenter Theil. *Palaeohatteria longicaudata* Cred. Z. Dtsch. Geol. Ges., 40: 490–558.

33. Cole, L. J. 1909. The destruction of birds at Niagara Falls. Auk, 26: 63–65.

34. Cuvier, G. 1834. Recherches sur les ossements fossiles. Vol. 3. 4th ed. 435 pp. Edmond D'Ocagne; Paris.

35. Dames, W. 1891. Ueber Vogelreste aus dem Saltholmskalk von Limhamn bei Malmö. Bihang till K. Svenska Vet. Akad. Handling., 16, afd. 4, 1: 3–12.

36. Dames, W. 1895. Die Plesiosaurier der süddeutschen Liasformation. Abh. Kgl. Preuss. Akad. Wiss. Berlin, 1895 (2): 1–83.

37. Darwin, C. 1839. Journal and remarks. *In* Narrative of the Surveying Voyages of His Majesty's Ships *Adventure* and *Beagle,* between the years 1826 and 1836, describing their Examination of the Southern Shores of South America, and of the *Beagle's* Circumnavigation of the Globe. Vol. III. London. [Later reprints entitled "Journal of Researches into the Natural History and Geology of the Countries visited during the voyage of H.M.S. *Beagle"*— German translation 1893.]

37a. Sievers, W., E. Deckert, and W. Kükenthal, eds. 1897. Nordamerika. Bibliographisches Institut; Leipzig and Vienna.

38. Deecke, W. 1913. Paläontologische Betrachtungen. IV: Über Fische. N. Jb. Min., etc., 1913: 69–92.

39. Deecke, W. 1915. Paläontologische Betrachtungen. VII: Über Crustaceen. N. Jb. Min., etc., 1915: 112–26.

40. Deecke, W. 1926. Über die Triasfische. Paläont. Z., 8: 184–198.

40a. Deecke, W. 1923. Die Fossilisation. vi + 216 pp. Gebrüder Borntraeger, Berlin.

41. Dietrich, W. O. 1924. *Elephas antiquus* Recki n.f. aus dem Diluvium Deutsch-Ostafrikas, nebst Bemerkungen über die stammesgeschichtlichen Veränderungen des Extremitätenskelettes der Proboscidier. (Wissenschaftliche Ergebnisse der Oldoway-Expedition I.) Arch. Biontol., 4(1): 1–80.

42. Ditmar, R. 1913. Sterben und Vergehen in der Natur in kolloid-chemischer Beleuchtung. Naturwiss. Wochenschr., N.F., 12 (50): 785–90.

43. Drevermann, F. 1910. Neues aus der Schausammlung. Ein fossiler Hai. Ber. Senckenberg. Naturf. Ges. Frankfurt a.M., 41: 191–93.

44. Eastman, C. R. 1900. Fossil lepidosteids from the Green River Shales of Wyoming. Bull. Mus. Comp. Zool. Harvard Coll., 36: 67–75.

44a. Eastman, C. R. 1917. Fossil Fishes in the Collection of the United States National Museum. Proc. U. S. Nat. Mus. 52: 235–304.

45. Edinger, T. 1927. Über einige fossile Gehirne. Paläont. Z., 9: 379–402.

46. Fabre, H. J. N.d. Die Befreiung der jungen Schmeissfliegen. Kosmos [incorrectly given by Weigelt as 1908: 203].

47. Filhol, H. 1876–77. Recherches sur les phosphorites du Quercy. Etude des fossiles qu'on y rencontre et spécialement des mammifères. Ann. Sci. Geol. (Paris), 7 (7): 220 pp. (1876): Ann. Sci. Geol. (Paris), 8 (1): 340 pp. (1877)

48. Fleming, J. H. 1908. The destruction of the Whistling Swans (*Olor columbianus*) at Niagara Falls. Auk, 25: 306–9.

49. Fraas, O. 1870. Die Fauna von Steinheim. Mit Rücksicht auf die miocenen

Säugethier- und Vogelreste des Steinheimer Beckens. 54 pp. E. Schweizer-bart'sche Verlagshandlung; Stuttgart.

50. Fraas, O. 1877. *Aetosaurus ferratus* Fr. Die gepanzerte Vogel-Echse aus dem Stubensandstein bei Stuttgart. Folio, Jh. Ver. Naturk. Württemberg, 33: 1-22.

51. Fraas, E. 1902. Die Meercrocodilier (Thalattosuchia) des oberen Jura unter specieller Berücksichtigung von *Dacosaurus* und *Geosaurus.* Palaeontographica, 49: 1-72.

52. Fraas, E. 1903. *Rana Danubina* H. v. Meyer var. *rara* O. Fraas aus dem Obermiocän von Steinheim. Jh. Ver. Vaterl. Naturk. Württemberg, 59: 105-10.

53. Frech, F. 1906. Über die Gründe des Aussterbens der vorzeitlichen Tierwelt. Arch. Rassen- u. Gesellschaftsbiol., 3: 469-98.

54. Freygang, J. 1923. Gliederung und Fossilgehalt des Kupferschiefers. Jb. Hall. Verb., 4: 183-92.

55. Freyberg, B. v. Oral communication.

56. Fuchs, T. 1895. Studien über Fucoiden und Hieroglyphen. Denkschr. Akad. Wiss. Wien, Math.-Naturw. Kl., 22: 369-448. [* Pl. VI, fig. 1.]

57. Führer durch das Museum der Kgl. Ungar. Geol. Reichsanst. Budapest (n.d.)

58. Gaillard, C. 1908. Les oiseaux des phosphorites du Quercy. Ann. Univ. Lyon, n.s. (Sci. méd.) 23: 1-178.

59. Gottlieb, H. 1914. Die Antiklinie der Wirbelsäule der Säugetiere. Morph. Jb., 49: 179-220.

60. Haardt, G.M. 1926. Through the deserts and jungles of Africa by motor. Nat. Geogr. Mag., 49 (1).

61. Hamilton, A. 1904. Note on remains of some of the extinct birds of New Zealand found near Ngapara. Proc. N. Zeal. Inst., 36: 474-77.

62. Hauff, B. 1921. Untersuchung der Fossilfundstätten von Holzmaden im Posidonienschiefer des oberen Lias Württembergs. Palaeontographica, 64: 1-42.

63. Haupt, O. 1925. Die Palaeohippiden der eocänen Süsswasserablagerungen von Messel bei Darmstadt. Abh. Grossherz.-Hess. Geol. Landesanst. Darmstadt, 6 (3): 1-159.

64. Heineke, E. 1907. Die Ganoiden und Teleostier des lithographischen Schiefers von Nusplingen. Geol. Paläont. Abh., N.F., 8: 159-214.

65. Heinroth, O. 1923. Die Flügel von *Archaeopteryx.* J. Ornith., 71: 277-83.

66. Hennig, E. 1915. Über dorsale Wirbelsäulenkrümmung fossiler Vertebraten. Cbl. Min., etc., 1915: 575-77.

67. Hennig, E. 1915. Eine neue Platte mit *Semionotus capensis.* Sitz.-Ber. Ges. Naturf. Freunde Berlin, 1915: 49-52.

68. Hennig, E. 1918. Ueber *Ptycholepis bollensis* Ag. Jh. Ver. Vaterl. Naturk. Württemberg, 74: 173-82.

69. Hinsche, G. Oral communication.

70. Hitchcock, E. 1858. Ichnology of New England. A Report on the Sandstone of the Connecticut Valley, especially its Fossil Footmarks. xii + 220 pp. William White; Boston.
Hitchcock, E. 1865. Supplement to the Ichnology of New England. A Report to the Government of Massachusetts in 1863. x + 96 pp. Wright and Potter; Boston.

71. Hoernes, R. 1911. Das Aussterben der Arten und Gattungen. Biol. Cbl., 31: 353–65, 385–94.
72. Huene, F. v. 1902. Übersicht über die Reptilien der Trias. Geol. Paläont. Abh., N.F., 6: 3-84. [*Fig. 32, p. 28.]
73. Huene, F. v. 1912. Die Cotylosaurier der Trias. Palaeontographica, 59: 69–102.
74. Huene, F. v. 1915. Beiträge zur Kenntnis einiger Saurischier der schwäbischen Trias. N. Jb. Min., etc., 1915: 1–27.
75. Huene, F. v. 1922. Die Ichthyosaurier des Lias und ihre Zusammenhänge. Monogr. Geol. Paläont., 1: viii + 114 pp. Gebrüder Borntraeger; Berlin.
76. Huene, F. v. 1925. [Weigelt's original reference cannot be found—possibly different journal or year.]
77. Huene, F. v. 1925. Ein neuer Pelycosaurier aus der unteren Permformation Sachsens. Geol. Paläont. Abh., N.F., 14: 215–64.
78. Hummel, K., and W. Wenz. 1924. Eine Maar-Ausfüllung mit obermiocaener Schneckenfauna bei Homburg a.d. Ohm im nördlichen Vogelsberg. Notizbl. Ver. Erdk. u. Hess. Geol. Landesanst. Darmstadt, (5) 6: 285–98.
79. Jaekel, O. 1889. Die Selachier aus dem oberen Muschelkalk Lothringens. Abh. Geol. Specialk. Elsass-Lothringen, 3: 275–332.
80. Jaekel, O. 1899. Über die Organisation der Petalodonten. Z. Dtsch. Geol. Ges., 51: 258–98. [*Pl. XIV and pl. XV, fig. 1.]
81. Jaekel, O. 1894. Die eocänen Selachier vom Monte Bolca. Ein Beitrag zur Morphogenie der Wirbelthiere. 176 pp. J. Springer; Berlin.
82. Janensch, W. 1906. Über *Archaeophis proavus* Mass., eine Schlange aus dem Eocän des Monte Bolca. Beitr. Geol. Paläont. Osterr.-Ung., 19: 1–33.
83. Johnston, W. A. 1922. Imbricated structure in river gravels. Amer. J. Sci., ser. 5, 4: 387–90.
84. Kaiser, E. 1926. Die Diamantenwüste Südwestafrikas. 2 vols. Dietrich Reimer A.G.; Berlin.
85. Kaudern, W. 1918. Über quartäre Fossilien aus Madagaskar. Zool. Jb., Abt. Syst., 41: 521–34.
86. Amalitzky, V. P. 1922. Diagnoses of the new forms of vertebrates and plants from the Upper Permian on North Dvina. Bull. Acad. Sci. St. Petersburg, (6) 16: 329–40.
87. Keith, A. 1925. When Malta was part of the Eur-African land bridge: A prehistoric big-game drive. Illustrated London News, 166 (February 28, 1925): 349–51.
88. Klähn, H. 1921. Die Ursache des Todes (nicht Aussterbens) tertiärer und pleistocäner Säuger in der mittleren Rheinebene. Z. Dtsch. Geol. Ges., B (Monatsber.), 73: 229. [Title of lecture only.]
89. Klähn, H. 1923. Die Beziehungen zwischen miocänen Sedimenten und den darin liegenden Landsäugetieren. Jb. Mitt. Oberrhein. Geol. Ver., N.F., 12: 152–58.
90. Klähn, H. 1924. Ueber einige säugetiereführende Vorkommnisse der Molasse Badens. N. Jb. Min., etc., Beil.-Bd. 50: 335–63.
91. Klähn, H. 1925. Die Säuger des badischen Miocäns. Palaeontographica, 66: 163–242.
92. Koken, E. 1887. Die Dinosaurier, Crocodiliden und Sauropterygier des norddeutschen Wealden. Geol. Paläont. Abh., 3: 311–419.

93. Kormos, T. 1911. Der pliozäne Knochenfund bei Polgardi. (Vorläufiger Bericht.) Földtani Közlöni, 41: 48–64 (Hungarian text), 171–89 (German text).

94. Kornhuber, A. 1873. Über einen neuen fossilen Saurier aus Lesina. Abh. K. K. Geol. Reichsanst. Wien, 5: 75–90.

Kornhuber, A. 1901. *Opetiosaurus bucchichi.* Eine neue fossile Eidechse aus der unteren Kreide von Lesina in Dalmatien. Abh. K. K. Geol. Reichsanst. Wien, 17: 1–24.

95. Kräusel, R. 1922. Die Nahrung von *Trachodon.* Paläont. Z., 4: 80.

96. Lambe, L. M. 1910. Palaeoniscid fishes from the Albert shales of New Brunswick. Contr. Can. Palaeont., 3, 5 (3): 7–35.

97. Langenhahn, A. 1905–9. Fauna und Flora des Rotliegenden in der Umgebung von Friedrichroda in Thüringen. Vol. I: 12 pp. (1905); vol. II: 2 pp. (1909).

98. Lankester, E. Ray. 1883. On the tusks of the fossil walrus, found in the Red Crag of Suffolk. Trans. Linn. Soc. Lond., (2) 2: 213–21.

99. Laube, G. C. 1903. Batrachier und Fischreste aus der Braunkohle von Skiritz bei Brüx. Lotos, 51: 106–14.

100. (Several special papers by J. Leidy and by E. D. Cope.)

101. Lucas, A. F. 1899. The fossil bison of North America. Proc. U. S. Nat. Mus., 21: 755–71.

101. Ludwig, R. 1877. Fossile Crocodiliden aus der Tertiärformation des Mainzer Beckens. Palaeontographica, Suppl. 3 (4–5): 1–54.

103. Lull, R. S. 1914. Fossil dolphin from California. Amer. J. Sci., (4) 37: 209–20.

104. Marshall, P. 1919. Occurrence of fossil moa-bones in the Lower Wanganui strata. Trans. Proc. N. Z. Inst., 51: 250–53.

105. Marck, W. v. d. 1863. Fossile Fische, Krebse und Pflanzen aus dem Plattenkalk der jüngsten Kreide in Westfalen. Palaeontographica, 11: 1–83.

106. Marsh, O. C. 1871. [Communication on some new reptiles and fishes from the Cretaceous and Tertiary.] Proc. Acad. Nat. Sci. Philadelphia, 1871: 103–5.

107. Meyer, H. v. 1845. Zur Fauna der Vorwelt. Fossile Saeugethiere, Voegel und Reptilien aus dem Molasse-Mergel von Oeningen. vi + 52 pp. S. Schmerber'sche Buchhandlung; Frankfurt a. M.

108. Meyer, H. v. 1856. Zur Fauna der Vorwelt. Saurier aus dem Kupferschiefer der Zechstein-Formation. vi + 28 pp. Verlag von Heinrich Keller; Frankfurt a. M.

109. Meyer, H. v. 1859. *Melosaurus uralensis* aus dem permischen System des westlichen Urals. Palaeontographica, 7: 90–98.

110. Meyer, H. v. 1860. Zur Fauna der Vorwelt. Reptilien aus dem lithographischen Schiefer in Deutschland und Frankreich. viii + 142 pp. Verlag von Heinrich Keller; Frankfurt a. M.

110a. Meyer, H. v. 1860. *Coluber* (*Tropidonotus?*) *atavus* aus der Braunkohle des Siebengebirges. Palaeontographica, 7: 232–40.

111. Miller, L. H. 1909. *Teratornis,* a new avian genus from Rancho la Brea. Bull. Dept. Geol., Univ. Calif., 5: 305–17.

112. Miller, L. H. 1909. *Pavo californicus,* a fossil peacock from the Quarternary asphalt-beds of Rancho la Brea. Bull. Dept. Geol., Univ. Calif., 5: 285–89.

113. Miller, L. H. 1911. Additions to the avifauna of the Pleistocene deposits at Fossil Lake, Oregon. Bull. Dept. Geol., Univ. Calif., 6: 79-87.

114. Miller, L. H. 1925. Avifauna of the McKittrick Pleistocene. Bull. Dept. Geol., Univ. Calif., 15: 307-26.

115. Moodie, R. L. 1923. Paleopathology: An Introduction to the Study of Ancient Evidences of Disease. 567 pp. University of Illinois Press; Urbana, Ill. [*p. 117, pl. 49].

116. Moos, A. 1926. Zur Bildung von Ablagerungen mit Landsäugetierresten in der süddeutschen Molasse. Geol. Rdsch., 17: 8-21.

117. Morgan, W. C., and M. C. Tallmon. 1904. A fossil egg from Arizona. Bull. Dept. Geol., Univ. Calif., 3: 403-10.

118. Nehring, A. 1890. Ueber Tundren und Steppen der Jetzt- und Vorzeit, mit besonderer Berücksichtigung ihrer Faunen. 257 pp. Berlin.

119. Neumayer, L. 1913. Zur vergleichenden Anatomie des Schädels eocäner und rezenter Siluriden. Palaeontographica, 59: 251-88.

119a. Newberry, J. S. 1890. The Paleozoic Fishes of North America. Monogr. U. S. Geol. Surv., 16: 1-340.

120. Nopsca, F. v. 1902. Ueber das Vorkommen der Dinosaurier von Szentpéterfalva. (Briefliche Mitteilung.) Z. Dtsch. Geol. Ges., 54: 34-39.

121. Nopsca, F. v. 1923. *Eidolosaurus* und *Pachyophis*, zwei neue Neocom-Reptilien. Palaeontographica 65: 96-154.

122. Obrutschew [= Obruchev], W. A. 1926. Geologie von Sibirien. Fortschr. Geol. Paläont., 15: xii + 572 pp.

123. Omeliansky, W. 1913. Cellulosegärung. *In* Handwörterbuch der Naturwissenschaften, vol. 4: 512-14. Gustav Fischer; Jena.

124. Osborn, H. F. 1910. The Age of Mammals in Europe, Asia and North America. xvii + 635 pp. Macmillan; New York.

125. Pander, C. H. 1860. Ueber die Saurodipterinen, Dendrodipterinen, Glyptolepiden und Cheirolepiden des devonischen Systems. ix + 90 pp. St. Petersburg.

126. Peterson, O. A. 1906. The Agate Spring fossil quarry. Ann. Carnegie Mus., 3: 487-94.

127. Pia, J. v. [Remarks made on the occasion of Weigelt's lecture at Stuttgart.]

128. Plieninger, F. 1895. *Campylognathus Zitteli*, ein neuer Flugsaurier aus dem oberen Lias Schwabens. Palaeontographica, 41: 193-222.

129. Plieninger, F. 1901. Beiträge zur Kenntnis der Flugsaurier. Palaeontographica, 48: 65-90. [*Pl. IV, Munich specimen of *"Pterodactylus" [Germanodactylus] kochi.]*

130. Pompeckj, J. F. 1914. Das Meer des Kupferschiefers. Pp. 444-94. *In* Pompeckj, J. F. (ed.), Branca-Festschrift. Gebrüder Borntraeger; Leipzig.

131. Porsild, M. P. 1902. Bidrag til en Skildring af Vegetationen paa øen Disko tilligemed spredte topografiske og zoologiske Iagttagelser. Meddelser om Grønland, 25: 91-239.

132. Reiss, W. 1883. Ueber eine fossile Säugethier-Fauna von Punin bei Riobamba in Ecuador. Die geologischen Verhältnisse der Fundstellen fossiler Säugethier-Knochen in Ecuador. Geol. Paläont. Abh., 1: 3-18.

133. Reis, O. M. 1888. Über *Belonostomus, Aspidorhynchus* und ihre Beziehungen zum lebenden *Lepidosteus*. Sitz.-Ber. Bayer. Akad. Wiss., II. Cl., 17 (1887): 151-77.

134. Richter, R. 1926. Flachseebeobachtungen zur Paläontologie und Geologie. XII–XIV. Senckenbergiana 8.

135. Rothpletz, A. 1909. Über die Einbettung der Ammoniten in die Solnhofener Schichten. Abh. Bayer. Akad. Wiss., II. Cl., 24: 311–38.

136. Sander, H. 1918. Mumifizierung und Radioaktivität. Naturw. Wochenschr., N.F., 17 (42).

137. Schaede, R. 1918. Über Schrumpfungs- und Kohäsions-Mechanismen. Kosmos, 1918 (1).

138. Schaffer, F. X. 1916. Grundzüge der allgemeinen Geologie. 492 pp. Leipzig and Vienna.

139. Schellwien, E. 1901. Über *Semionotus*. Schriften Phys.-Ökonom. Ges. Königsberg i. Pr., 42: 1–33.

140. Scheunert, A. 1915. Verdauung. *In* Handwörterbuch der Naturwissenschaften, 10: 230–43. Gustav Fischer; Jena.

141. Schlosser, M. 1921. Die Hipparionenfauna von Veles in Mazedonien. Abh. Bayer. Akad. Wiss., math.-phys. Kl., 29 (4): 55 pp.

142. Schmidt, M. 1907. Labyrinthodontenreste aus dem Hauptkonglomerat von Altensteig im württembergischen Schwarzwald. Mitt. Geol. Abt., Kgl. Württemb. Statist. Landesamtes, 2: 10 pp.

143. Schmidt, M. 1921. *Hybodus hauffianus* und die Belemnitenschlachtfelder. Jh. Vaterl. Ver. Naturk. Württemberg, 77: 103–7.

144. Schmut, H. 1924. Ein untermiocäner Reptilienschädel-Ausguss. Cbl. Min., etc., 1924: 117–19.

145. Schubert, R. J. 1905. Die Fischotolithen des österreichisch-ungarischen Tertiärs. II. Macruriden und Beryciden. Jb. Geol. Reichsanst., 55: 613–38.

146. Schütze, E. 1901. Beiträge zur Kenntnis der triassischen Koniferengattungen: *Pagiophyllum, Voltzia* und *Widdringtonites*. Jh. Ver. Vaterl. Naturk. Württemberg, 57: 240–74.

147. Schroeder, H. C. 1918. Eocäne Säugetierreste aus Nord- und Mitteldeutschland. Jb. Preuss. Geol. Landesanst., 37: 164–95.

148. Schwarz, E. 1913: Über einen Schädel von *Palhyaena hipparionum* (Gervais) nebst Bemerkungen über die systematische Stellung von *Ictitherium* und *Palhyaena*. Arch. Naturgesch., 78 (A, 11): 69–74.

149. Scott, W. B. 1913. A History of the Land Mammals in the Western Hemisphere. xiv + 693 pp. Macmillan; New York.

150. Seidlitz, W. v. 1917. Ueber ein Krokodil aus den oligocänen Braunkohlenschichten von Camberg a. Saale. Jb. Preuss. Geol. Landesanst., 38: 347–67. [*Pl. 22, fig. I.]

151. Semon, R. 1893. Reisebericht und Plan des Werkes. Verbreitung, Lebensverhältnisse und Fortpflanzung des *Ceratodus Forsteri* (pp. 11–28). Die äussere Entwicklung des *Ceratodus Forsteri* (pp. 29–50). *In* Semon, R. (ed.), Zoologische Forschungsreisen in Australien und dem Malayischen Archipel, vol. 4. Gustav Fischer; Jena.

152. Shepard, C. N. 1867. On the supposed tadpole nests, or imprints made by *Batrachoides nidificans* (Hitchcock) in the Red Shale of the New Red Sandstone of South Hadley, Mass. Amer. J. Sci., ser. 2, 43: 99–104.

153. Soergel, W. 1912. Das Aussterben diluvialer Säugetiere und die Jagd des diluvialen Menschen. 81 pp. Gustav Fischer; Jena.

154. Sokolow, W. 1897. [Quelques données concernant le changement périod-

ique de la salure de l'eau dans le liman du Boug.] Izvest. Géol. Kom. St. Petersbourg, 16(4): 145–54 (in Russian with French abstract).

155. Stappenbeck. Oral communication.

156. Stensiö. Oral communication, 1926.

157. Stirling, E. C. 1900. The physical features of Lake Callabonna. Mem. Roy. Soc. S. Austr., 1: i–xv.

Stirling, E. C., and A. H. G. Zietz. 1900. Fossil remains of Lake Callabonna. Part I. Description of the manus and pes of *Diprotodon australis,* Owen. Mem. Roy. Soc. S. Austr., 1: 1–40.

Stirling, E. C., and A. H. G. Zietz. 1900. *Genyornis newtoni,* a new genus and species of fossil struthious bird. Mem. Roy. Soc. S. Austr., 1: 41–80.

158. Stromer [von Reichenbach], E. 1909–12. Lehrbuch der Palaeozoologie. 2 vols. B. G. Teubner; Leipzig.

159. Stromer, E. 1926. Reste Land- und Süsswasser-bewohnender Wirbeltiere aus den Diamantfeldern Deutsch-Südwestafrikas. pp. 107–53. *In* Kaiser, E., Die Diamantenwüste Südwest-Afrikas, vol. II. Berlin.

160. Studer, T. 1896. Die Säugetierreste aus den marinen Molasseablagerungen von Brüttelen. Abh. Schweizer. Paläont. Ges., 12: 47 pp.

161. Swinnerton, H. H. 1925. A new catopterid fish from the Keuper of Nottingham. Quart. J. Geol. Soc. Lond., 81: 87–99.

162. Thévenin, A. 1903. Etude géologique de la bordure sud-ouest du massif central. Bull. Carte Géol. France, 14: 153–354.

163. Thienemann, A. 1918. Lebensgemeinschaft und Lebensraum. Naturwiss. Wochenschr., N.F., 17: 281–90. [* P. 288.]

164. Thomas, E. Written communication.

165. Vollrath, P. 1923. *Ceratodus elegans* n. sp. aus dem Stubensandstein. Jb. Mitt. Oberrhein. Geol. Ver., 12: 158–63.

166. Volz, W. 1902. *Proneusticosaurus,* eine neue Sauropterygier-Gattung aus dem untersten Muschelkalk Oberschlesiens. Palaeontographica, 49: 121–62.

167. Wagner, J. A. 1861. Neue Beiträge zur Kenntniss der urweltlichen Fauna des lithographischen Schiefers. Zweite Abtheilung. Schildkröten und Saurier. Abh. Kgl. Bayer. Akad. Wiss., II. Cl., 9: 65–124.

168. Waibel, L. 1921. Urwald, Veld und Wüste. Breslau.

169 Walther, J. 1904. Die Fauna der Solnhofener Plattenkalke. Bionomisch betrachtet. Denkschr. Med.-Naturw. Ges. Jena, 11: 133–214.

170. Walther, J. 1927. Allgemeine Palaeontologie. xv + 809 pp. Gebrüder Borntraeger; Berlin. [* Chapters II and III.]

171. Walther, J. 1893–94. Einleitung in die Geologie als historische Wissenschaft. Gustav Fischer; Jena. [* Chapter III, pp. 268ff.]

172. Weigelt, J. 1919. Gliederung und Faunengehalt im Unteren Culm des Oberharzes. Jb. Preuss. Geol. Landesanst., 37: 157–271.

172a. Weigelt, J. 1923. Angewandte Geologie und Paläontologie der Flachseegesteine und das Erzlager von Salzgitter. Fortschr. Geol. Paläont., 4: iii + 128 pp.

173. Wepfer, E. 1923. Der Buntsandstein des badischen Schwarzwaldes und seine Labyrinthodonten. Monogr. Geol. Paläont., (2) 1: viii + 101 pp. Gebrüder Borntraeger; Berlin.

174. Wettstein, A. 1886. Über die Fischfauna des tertiären Glarnerschiefers. Abh. Schweiz. Paläont. Ges., 13: 103 pp.

178   REFERENCES

175.   Wigand, P. 1919. Zur Frage des Zusammenhanges zwischen Mumifikation und Radioaktivität. Naturwiss. Wochenschr., N.F. 18 (6).
176.   Wiman, C. 1913. Über die paläontologische Bedeutung des Massensterbens unter den Tieren. Paläont. Z., 1: 145–54.
177.   Wiman, C. 1920. Some reptiles from the Niobrara group in Kansas. Bull. Geol. Instn. Uppsala, 18: 9–18.
178.   Wiman, C. 1923. Ueber *Dorygnathus* und andere Flugsaurier. Bull. Geol. Instn. Uppsala, 19: 23–54.
179.   Wiman, C. 1923. Massentod von Vögeln in Niagara. Paläont. Z., 5: 105–7.
180.   Winge, H. 1906. Jordfunde og nulevende hovdyr (Ungulata) fra Lagoa Santa, Minas Geraes, Brasilien. Med udsigt over hovdyrens inyrdes-slaegtskab. E Museo Lundii, 3 (1): 1–239. Copenhagen.
181.   Wöhlbier. 1910. Mansfelder Fischabdrücke. Sonntagsblatt, supplement to Eislebener Zeitung, no. 11.
181.   Woodward, A. S. 1907. On a reconstructed skeleton of *Diprotodon* in the British Museum (Natural History). Geol. Mag., (5) 4: 337–39.
183.   Woodward, A.S. 1902–12. Fossil Fishes of the English Chalk. Monogr. Palaeontogr. Soc., 61–64.
184.   Wüst, E. 1922. Beiträge zur Kenntnis der diluvialen Nashorner Europas. Cbl. Min., etc., 1922: 641–56 and 680–88.
185.   Zittel, K. A. v. 1918. Grundzüge der Paläontologie. (Paläozoologie.) II. Abt. Vertebrata. R. Oldenbourg; Munich. [* Fig. 474, p. 355.]

Index